# AN ATLAS OF
# MAMMALIAN CHROMOSOMES

VOLUME 5

AF838402

# AN ATLAS OF MAMMALIAN CHROMOSOMES

VOLUME 5

T. C. HSU

Section of Cytology, Department of Biology, The University of Texas M. D. Anderson Hospital and Tumor Institute, Houston, Texas

KURT BENIRSCHKE

Department of Obstetrics & Gynecology, School of Medicine, University of California, San Diego, La Jolla, California

SPRINGER-VERLAG
BERLIN • HEIDELBERG • NEW YORK
1971

All rights reserved, especially that of translation into foreign languages.
It is also forbidden to reproduce this book, either whole or in part, by photomechanical means (photostat, microfilm, and/or microcard) or by other procedure without written permission from Springer-Verlag New York Inc.
Library of Congress Catalog Card Number 67-19307
Printed in the United States of America
© 1971 by Springer-Verlag New York Inc.

ISBN 3-540-05280-1 Springer-Verlag Berlin Heidelberg New York
ISBN 0-387-05280-1 Springer-Verlag New York Heidelberg Berlin

## Introduction

Since the inception of the Atlas of Mammalian Chromosomes, it is a surprise to us that we have finished five volumes with karyotypes of nearly 250 mammalian species. We acknowledge the fact that there are many imperfections in this series, but we also hope that these volumes have been useful to various investigators.

Springer-Verlag has made a binder which will accommodate the material of all five volumes. We also revised our index to facilitate easy reference. If the material is arranged in the binder according to the Cumulative Contents, we believe one will find it convenient to consult with this book.

Volume 5, as in the previous volumes, contains a Cumulative Index, Cumulative Table of Contents, as well as an individual Index and Table of Contents for Volume 5. Previous Indices and Tables of Contents should be discarded.

October, 1970

T. C. Hsu
Kurt Benirschke

## Instructions

A special vinyl binder is available from the publisher which will accommodate the material of the first five volumes. Twenty dividers have been provided with printed tabs. These dividers are intended to separate the Orders and the Index. It is suggested that the dividers be placed in the following sequence of Orders, which sequence will be followed in the future when presenting karyotypes in new Orders:

Monotremata
Marsupialia
Insectivora
Dermoptera
Chiroptera
Edentata
Tubulidentata
Lagomorpha
Rodentia
Sirenia
Pholidota
Hyracoidea
Proboscidea
Carnivora
Pinnipedia
Cetacea
Perissodactyla
Artiodactyla
Primates
Index

These additional features should make it more convenient to use the Atlas in the future.

1. Folios should be arranged according to the Cumulative Table of Contents of Volumes 1-5, which is included in Volume 5. *If the Vinyl Binder is used, contents may be separated according to the taxonomic orders, each of which has an index plate for convenience.*
2. Use the Cumulative Index (Volumes 1-5) to find whether a particular species is included. If so, find the order and family of that particular species (abbreviations in parentheses) and follow the folio numbers.
3. The new Bibliographical References should, as in the past, be incorporated into the previous folios. These new references are printed on only one side of their respective sheets and are arranged in accordance with the order of this Atlas and folio numbers. They should be cut apart and pasted into the respective folio as indicated.

# Contents, Volume 5

# Cumulative Contents (Volumes 1 to 5)

## New References for Previous Volumes

In Volume 5 there are 50 new species of various orders. These should be integrated with the Folios of Volumes 1, 2, 3 and 4. A new cumulative index is supplied, hence the old index should be discarded.

In addition to the new Folios, 5 pages of references for the Folios of Volumes 1, 2, 3 and 4 have been prepared to keep the Atlas up-to-date and useful. These references are arranged in such a fashion that they can be cut out and pasted into the previous Folios in proper sequence.

MARSUPIALIA

Vol. 1, Folio 1

12) Brinkley, B.R. and Humphrey, R.M.: Evidence for subchromatid organization in marsupial chromosomes. J. Cell Biol. 42:827, 1969.

INSECTIVORA

Vol. 2, Folio 52

10) Gropp, A., Citoler, P. and Geisler, M.: Karyotypvariation und Heterochromatinmuster bei Igeln (*Erinaceus* und *Hemiechinus*). Chromosoma 27: 288, 1969.

Vol. 2, Folio 53

9) Gropp, A., Citoler, P. and Geisler, M.: Karyotypvariation und Heterochromatinmuster bei Igeln (*Erinaceus* und *Hemiechinus*). Chromosoma 27: 288, 1969.

EDENTATA

Vol. 4, Folio 160

3) Corin-Frederic, J.: Les formules gonosomiques dites aberrantes chez les Mammifères Euthériens. Exemple particulier du paresseux *Choloepus hoffmanni* Peters (Edente, Xenarthre, famille des *Bradypodidae*). Chromosoma 27:268, 1969.

LAGOMORPHA

Vol. 1, Folio 8

18) Shaver, E.L. and Carr, D.H.: The chromosome complement of rabbit blastocysts in relation to the time of mating and ovulation. Canad. J. Genet. Cytol. 11:287, 1969.

RODENTIA

Vol. 2, Folio 73

8) Fernández, L.R. and Spotobno, A.: Heteromorphism in chromosome pair No. 1 of *Cavia porcellus*. Arch. Biol. y Medic. Exper. 5:81, 1969.

9) Jagiello, G.M.: Some cytologic aspects of meiosis in female guinea pig. Chromosoma 27:95, 1969.

RODENTIA

Vol. 1, Folio 13

15) Deaven, L.L. and Stubblefield, E.: Segregation of chromosomal DNA in Chinese hamster fibroblasts in vitro. Exp. Cell Res. 55:132, 1969.

16) Dewey, W.C. and Miller, H.H.: X-ray induction of chromatid exchanges in mitotic and $G_1$ Chinese hamster cells pretreated with colcemid. Exp. Cell Res. 57:63, 1969.

17) Sonnenschein, C., Roberts, D.W. and Yerganian, G.: Karyotypic and enzymatic characteristics of a somatic hybrid cell line originating from dwarf hamsters. Genetics 62:379, 1969.

18) Yerganian, G., Nell, M.A., Cho, S.S., Hayford, A.H. and Ho, T.: Virus-associated gain and loss of proliferative and neoplastic properties of normal and virus-transformed diploid cell lines. Nat. Cancer Inst. Monogr. No. 29: 241, 1968.

Vol. 1, Folio 14

15) Sasaki, M. and Kamada, T.: A phenotypically normal female golden hamster with sex-chromosome anomaly. Jap. J. Genet. 44:11, 1969.

16) Raicu, P., Ionescu-Varo, M. and Duna, D.: Interspecific crosses between the Rumanian and Syrian hamster. Cytogenetic and histologic studies. J. Hered. 60:149, 1969.

Vol. 1, Folio 17

21) Fraccaro, M., Hansson, K., Hultén, M., Lindsten, J. and Tiepolo, L.: Heterochromatin in preimplantation mouse embryos. Exp. Cell Res. 55:427, 1969.

22) Lyon, M.F.: A true hermaphrodite mouse presumed to be an XO/XY mosaic. Cytogenetics 8:326, 1969.

23) Léonard, A. and Deknudt, Gh.: Etude cytologique d'une translocation chromosome Y-autosome chez la souris. Experientia 25:876, 1969.

24) Léonard, A. and Deknudt, Gh.: Etude d'une translocation de type Robertsonien chez les souris de race AKR. Acta Zool. Path. Antv. 48:43, 1969.

Vol. 2, Folio 71

8) Sen, Y.H.: Karyotypes of Malayan rats (Rodentia-Muridae, genus Rattus Fischer). Chromosoma 27:245, 1969.

RODENTIA

Vol. 1, Folio 18

19) Hori, S.H. and Sasaki, M.: Glucose 6-phosphate dehydrogenase isoenzyme patterns and chromosomes in primary liver tumors of the rat. Cancer Res. 29:880, 1969.

20) Sen, Y.H.: Karyotypes of Malayan rats (Rodentia-Muridae, genus Rattus Fischer). Chromosoma 27:245, 1969.

21) Iversen, J.G.: Phytohemagglutinin response of recirculating and non-recirculating rat lymphocytes. Exp. Cell Res. 56:219, 1969.

CARNIVORA

Vol. 1, Folio 20

16) Ford, L.: Identification and chromomeric interpretation of pachytene bivalents from Canis familiaris. Canad. J. Genet. Cytol. 11:389, 1969.

17) Clough, E., Pyle, R.L., Hare, W.C.D., Kelly, D.F. and Patterson, D.F.: An XXY sex-chromosome constitution in a dog with testicular hypoplasia and congenital heart disease. Cytogenetics 9:71, 1970.

Vol. 2, Folio 78

3) Benirschke, K.: Zoos and the pathologist - a two way street or Cytogenetics on zoo animals. Acta Zool. Path. Antv. 48:29, 1969.

PERISSODACTYLA

Vol. 1, Folio 34

9) Pirtle, E.C. and Woods, L.K.: Observations of chromosomes of the horse. Mammalian Chromosomes Newsletter 9:3, 1968.

ARTIODACTYLA

Vol. 3, Folio 137

5) Koulischer, L.: Concept of cellular clonal evolution of karyotypes applied to evolution of species. Acta Zool. Path. Antv. 48:21, 1969.

Vol. 3, Folio 138

4) Bhambhani, R. and Kuspira, J.: The somatic karyotypes of American bison and domestic cattle. Canad. J. Genet. Cytol. 11:243, 1969.

ARTIODACTYLA

Vol. 1, Folio 44

25) Amrud, J.: Centric fusion of chromosomes in Norwegian red cattle (NRF). Hereditas 62:293, 1969.

26) Bhambhani, R. and Kuspira, J.: The somatic karyotypes of American bison and domestic cattle. Canad. J. Genet. Cytol. 11:243, 1969.

27) Gustavsson, J.: Cytogenetics, distribution and phenotypic effects of a translocation in Swedish cattle. Hereditas 63:68, 1969.

Vol. 3, Folio 140

6) Koulischer, L.: Concept of cellular clonal evolution of karyotypes applied to evolution of species. Acta Zool. Path. Antv. 48:21, 1969.

Vol. 1, Folio 45

19) Bruère, A.N.: Male sterility and an autosomal translocation in Romney sheep. Cytogenetics 8:209, 1969.

20) Bruère, A.N., Marshall, R.B. and Ward, D.P.J.: Testicular hypoplasia and XXY sex chromosome complement in two rams: the ovine counterpart of Klinefelter's syndrome in man. J. Reprod. Fert. 19:103, 1969.

21) Koulischer, L.: Concept of cellular clonal evolution of karyotypes applied to evolution of species. Acta Zool. Path. Antv. 48:21, 1969.

22) Jönsson, G. and Gustavsson, I.: Blood cell chimerism in one of three triplet lambs. J. Hered. 60:175, 1969.

Vol. 1, Folio 39

4) Lawrence, H.L.: Wild boars of the Appalachians. Natural History 78:46, 1969.

PRIMATES

Vol. 2, Folio 98

5) Egozcue, J., Perkins, E.M., Hagemenas, F. and Ford, D.M.: The chromosomes of some Platyrrhini (*Callicebus*, *Ateles* and *Saimiri*). Folia primat. 11:17, 1969.

Vol. 1, Folio 48

10) Chiarelli, B.: The phylogeny of primates from a karyological point of view. Acta Zool. Path. Antv. 48:11, 1969.

11) Markarian, D.S., Matchavariani, M.G. and Avjian, M.V.: The normal karyotype of the green monkey (Cercopithecus aethiops). Genetika 5:143, 1969.

Vol. 1, Folio 50

11) Al-Aish, M.S.: Human chromosome morphology. I. Studies on normal chromosome characterization, classification and karyotyping. Canad. J. Genet. Cytol. 11:370, 1969.

Vol. 3, Folio 147

11) Egozcue, J.: Meiosis in five Macaca species. Folia primat. 11:1, 1969.

Vol. 3, Folio 146

4) Elliot, O.S., Borgaonkar, D.S. and Wong, M.: Chromosome number as related to general biological affinities of tree shrews. Mammalian Chromosomes Newsletter 9:51, 1968.

5) Elliot, O.S., Wong, M. and Borgaonkar, D.S.: Karyological study of Tupaia from Thailand. J. Hered. 60:153, 1969.

Vol. 2, Folio 97

7) Elliot, O.S., Borgaonkar, D.S. and Wong, M.: Chromosome number as related to general biological affinities of tree shrews. Mammalian Chromosomes Newsletter 9:51, 1968.

8) Elliot, O.S., Wong, M. and Borgaonkar, D.S.: Karyological study of Tupaia from Thailand. J. Hered. 60:153, 1969.

# Order: MARSUPIALIA

## Family: DIDELPHIDAE

### *Didelphis marsupialis*

$2n = 22$

Order: MARSUPIALIA    Family: DIDELPHIDAE

*Didelphis marsupialis*

2n=22

AUTOSOMES: 20 Acrocentrics

SEX CHROMOSOMES: X Acrocentric
Y Acrocentric

The American opossum, *Didelphis marsupialis virginiana* Kerr (Folio 51), possesses a diploid number of 22 but 12 autosomes and the X are biarmed. This feature is consistent among all specimens collected in USA and in a portion of Mexico. However, *D. marsupialis* specimens collected in South America and Central America displayed no biarmed elements as depicted here. An extensive study made by Dr. Alfred L. Gardner (personal communication) showed that in Mexico both forms can be found, but the two are apparently isolated by ecological barriers. According to Gardner's opinion, the two should be considered separate species. At any rate, the karyotype in Folio 51 should have its subspecies name *virginiana* affixed.

The karyotypes presented here are gifts of Dr. A. L. Gardner. The male specimen bears specimen number ALG 11385 and is deposited in the Museum of Zoology, Louisiana State University, Baton Rouge, Louisiana, USA. The female specimen was collected 5 Km W. by road from El Naranjo, San Luis Potosí, Mexico. Bone marrow was used for the male and skin culture was used for the female specimens.

Didelphis marsupialis

2n=22

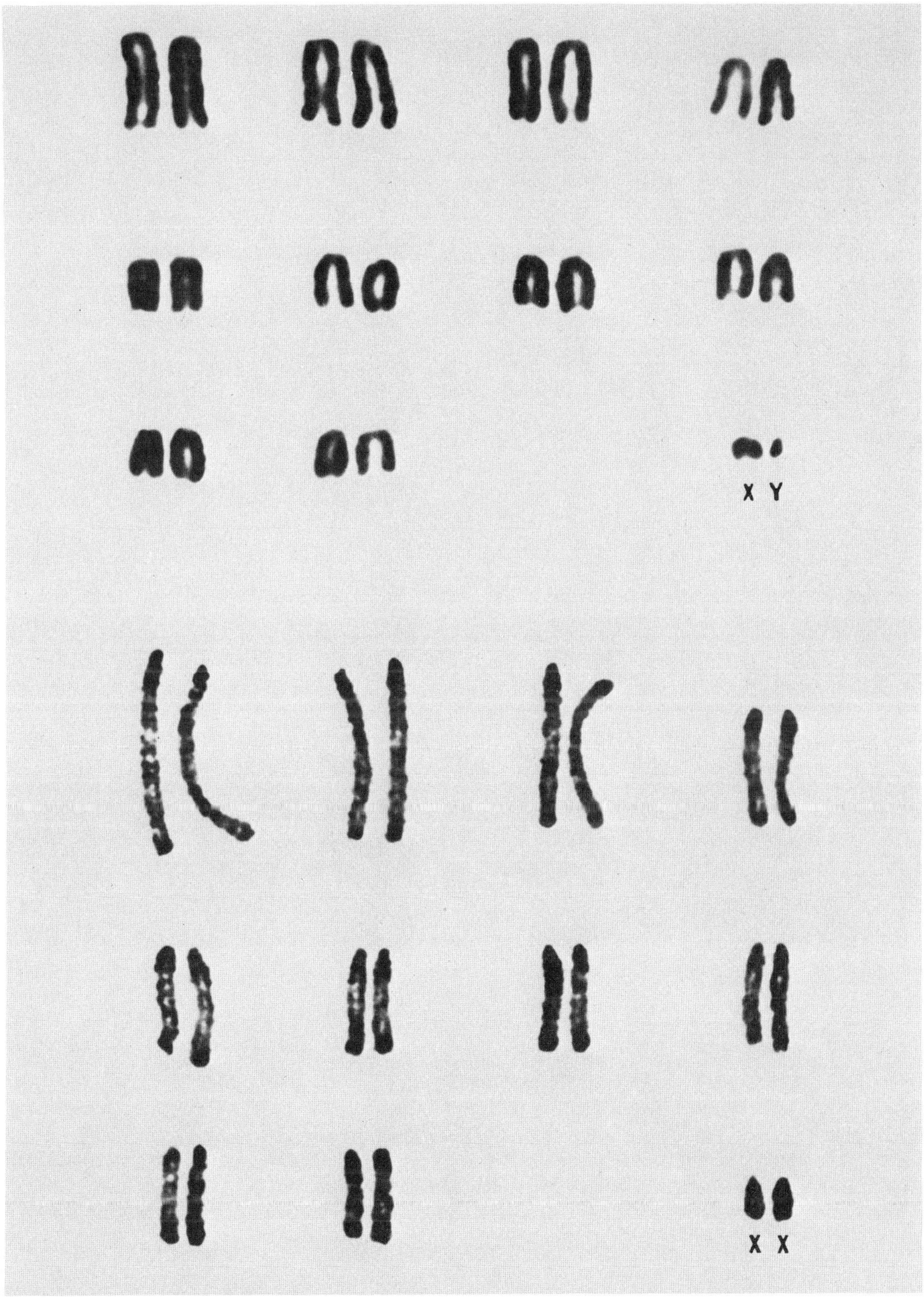

# Order: MARSUPIALIA

## Family: DIDELPHIDAE

### *Marmosa alstoni* (Alston's opossum)
### 2n = 14

Order: MARSUPIALIA Family: DIDELPHIDAE

*Marmosa alstoni* (Alston's opossum)

2n=14

AUTOSOMES: 8 Metacentrics and submetacentrics
2 Subtelocentrics
2 Acrocentrics

SEX CHROMOSOMES: X Acrocentric
Y Acrocentric

The X chromosome can either be called subtelocentric or acrocentric. It is about as long as, or slightly longer than, the subtelocentric autosome. This feature can be used to distinguish from the karyotype of *M. murina* (Folio 203 ), whose X is shorter than the subtelocentric autosome.

The karyotypes presented here are gifts of Dr. James L. Patton. The specimens were originally collected by Dr. Alfred L. Gardner, at ca 2km N.W. Santa Ana, Prov. San José, Costa Rica. They bear the voucher specimen numbers ♂12638 and ♀12637, Museum of Zoology, Louisiana State University, Baton Rouge, Louisiana, USA. Bone marrows were used for cytological preparations.

*Marmosa alstoni* (Alston's opossum)

2n=14

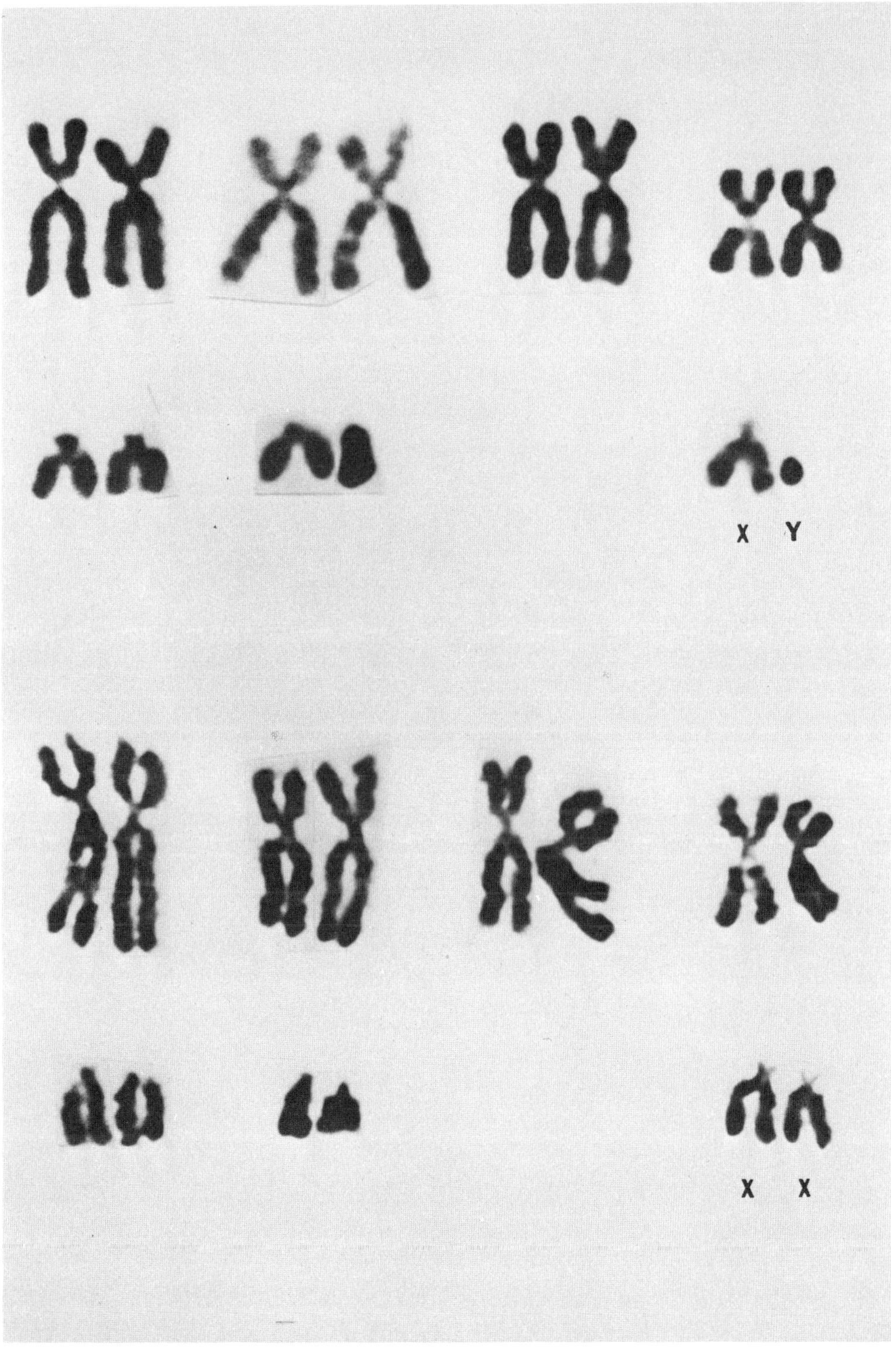

# Order: MARSUPIALIA

## Family: DIDELPHIDAE

***Marmosa murina*** (Mouse opossum)
$2n = 14$

Order: MARSUPIALIA Family: DIDELPHIDAE

<u>Marmosa murina</u> (Mouse opossum)

2n=14

AUTOSOMES: 8 Metacentrics and submetacentrics
2 Subtelocentrics
2 Acrocentrics

SEX CHROMOSOMES: X Submetacentric
Y Acrocentric

The X chromosome may be considered acrocentric. It is the smallest of the complement besides the Y. This is probably the only feature which can be distinguished from the karyotype of <u>M. alstoni</u> (Folio 202) whose X chromosome appears to be as long as the subtelocentric pair. The karyotypes of <u>Marmosa</u> are also similar to that of <u>Caluromys</u> (Folio 101).

These karyotypes are gifts of Dr. James L. Patton. The specimens were collected from Balta, Rio Curanja, Depto Loreto, Peru (El. 300 meters). The specimens are deposited in the Museum of Zoology, University of California at Berkeley, California, USA, bearing voucher numbers ♂136370 and ♀136368. Bone marrow was used for cytological preparations.

Order: MARSUPIALIA Family: DIDELPHIDAE

*Marmosa murina* (Mouse opossum)

2n=14

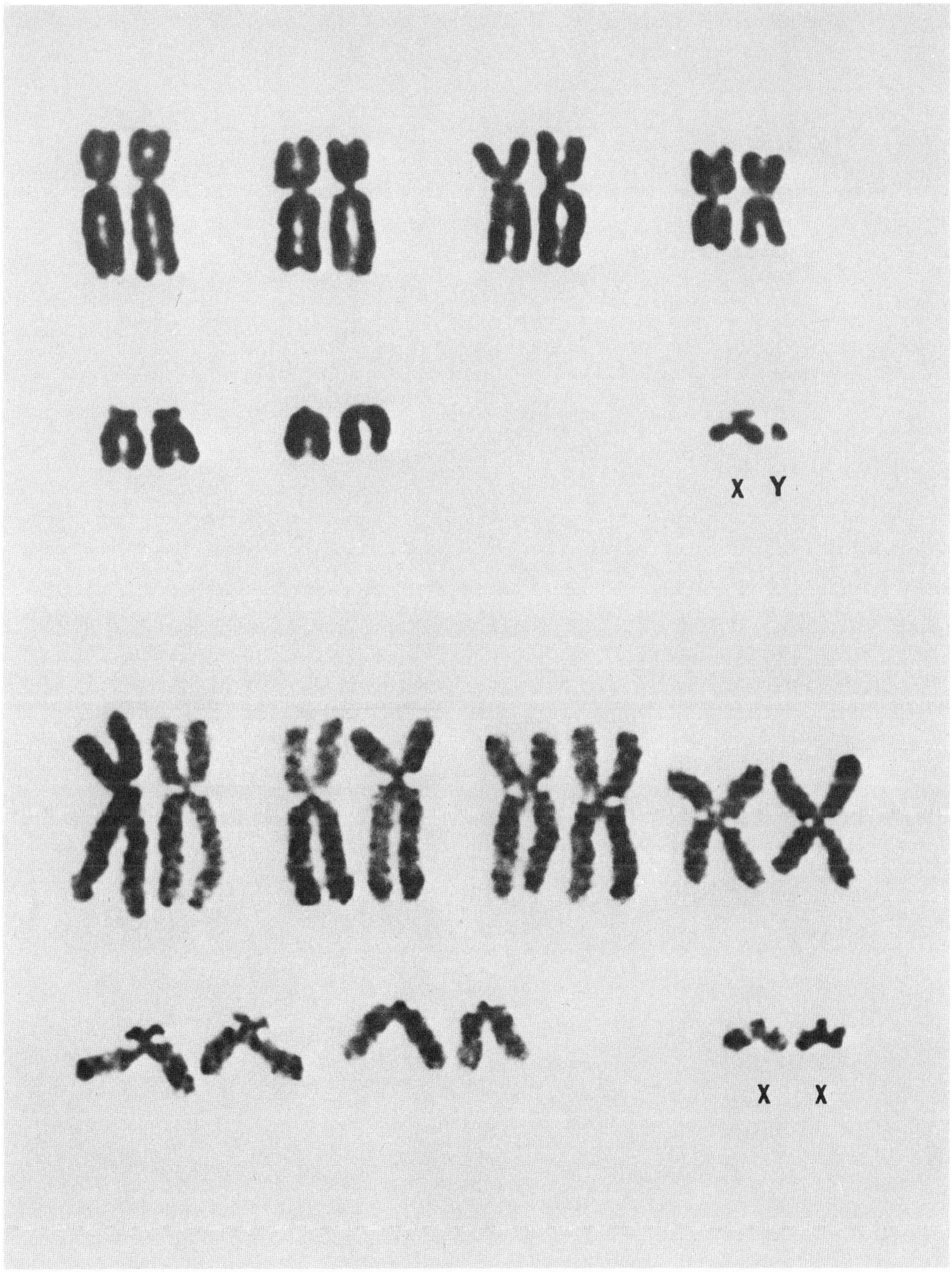

# Order: INSECTIVORA

## Family: SORICIDAE

*Neomys fodiens* (Old world water shrew)

2n = 52

Volume 5, Folio 204, 1971

Order: INSECTIVORA Family: SORICIDAE

Neomys fodiens (Old world water shrew)

2n=52

AUTOSOMES: 44 Metacentrics, submetacentrics and subtelocentrics
6 Acrocentrics

SEX CHROMOSOMES: X Submetacentric
Y Submetacentric

The karyotypes were prepared and kindly donated by Dr. K. Fredga, Lund, Sweden. Five males and two females from Southern Sweden were studied. Various tissues, including lymphocytes and direct spleen preparations, gave similar results. The male karyotype displayed came from a lung culture, that of the female, from a skin culture. Fredga and Levan give a detailed karyotype description and arm ratios. The X constitutes 5.6% of the haploid set. Satellite association of many acrocentrics is frequent.

REFERENCES:

1) Bovey, R.: Les chromosomes de Chiroptères et des Insectivores. Rev. Suisse Zool. 56:371, 1949.

2) Fredga, K. and Levan, A.: The chromosomes of the European water shrew (Neomys fodiens). Hereditas 62:348, 1969.

Neomys fodiens (Old world water shrew)

2n=52

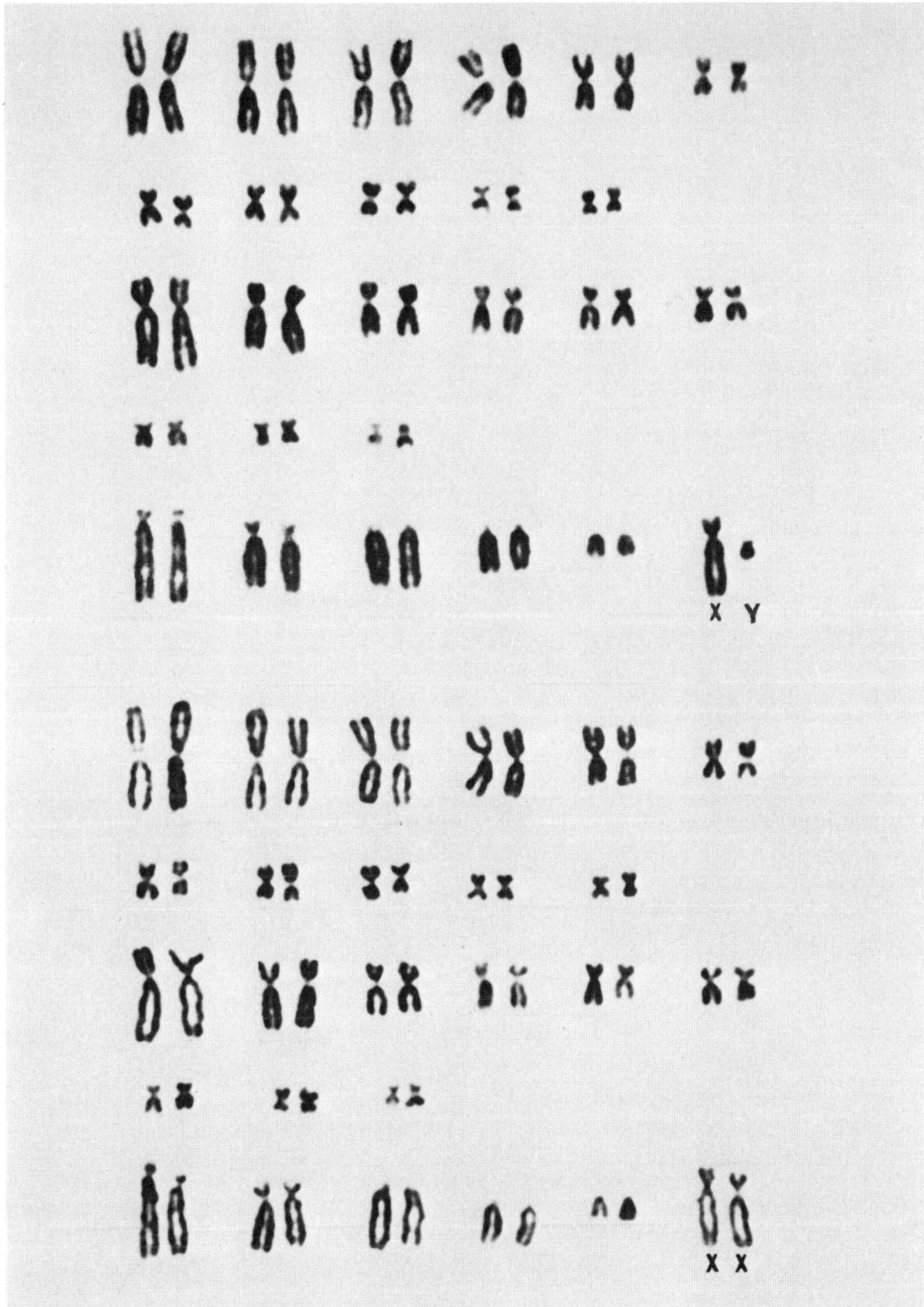

# Order: INSECTIVORA

## Family: SORICIDAE

*Sorex caecutiens* (Masked shrew; Laxmann's shrew)
2n = 42

Order: INSECTIVORA Family: SORICIDAE

*Sorex caecutiens* (Masked shrew; Laxmann's shrew)

2n=42

AUTOSOMES: 20 Metacentrics and submetacentrics
20 Acrocentrics

SEX CHROMOSOMES: X Acrocentric
Y Acrocentric

These karyotypes were kindly donated by Dr. K. Fredga, Lund, Sweden, and are arranged according to Fredga's system. Some of the smaller acrocentrics have pronounced short arms and could be considered submetacentrics by others. The smallest autosomes have satellites and engage in satellite association. One male and two females from Sweden were studied and the karyotypes were made from direct spleen preparations.

REFERENCES:


1) Skaren, U. and Halkka, O.: The karyotype of *Sorex caecutiens* Laxmann. Hereditas 54:376, 1966.

2) Fredga, K.: Chromosomes of the masked shrew (*Sorex caecutiens* Laxm.) Hereditas 60:269, 1968.

Order: INSECTIVORA

Family: SORICIDAE

*Sorex caecutiens* (Masked shrew; Laxmann's shrew)

2n=42

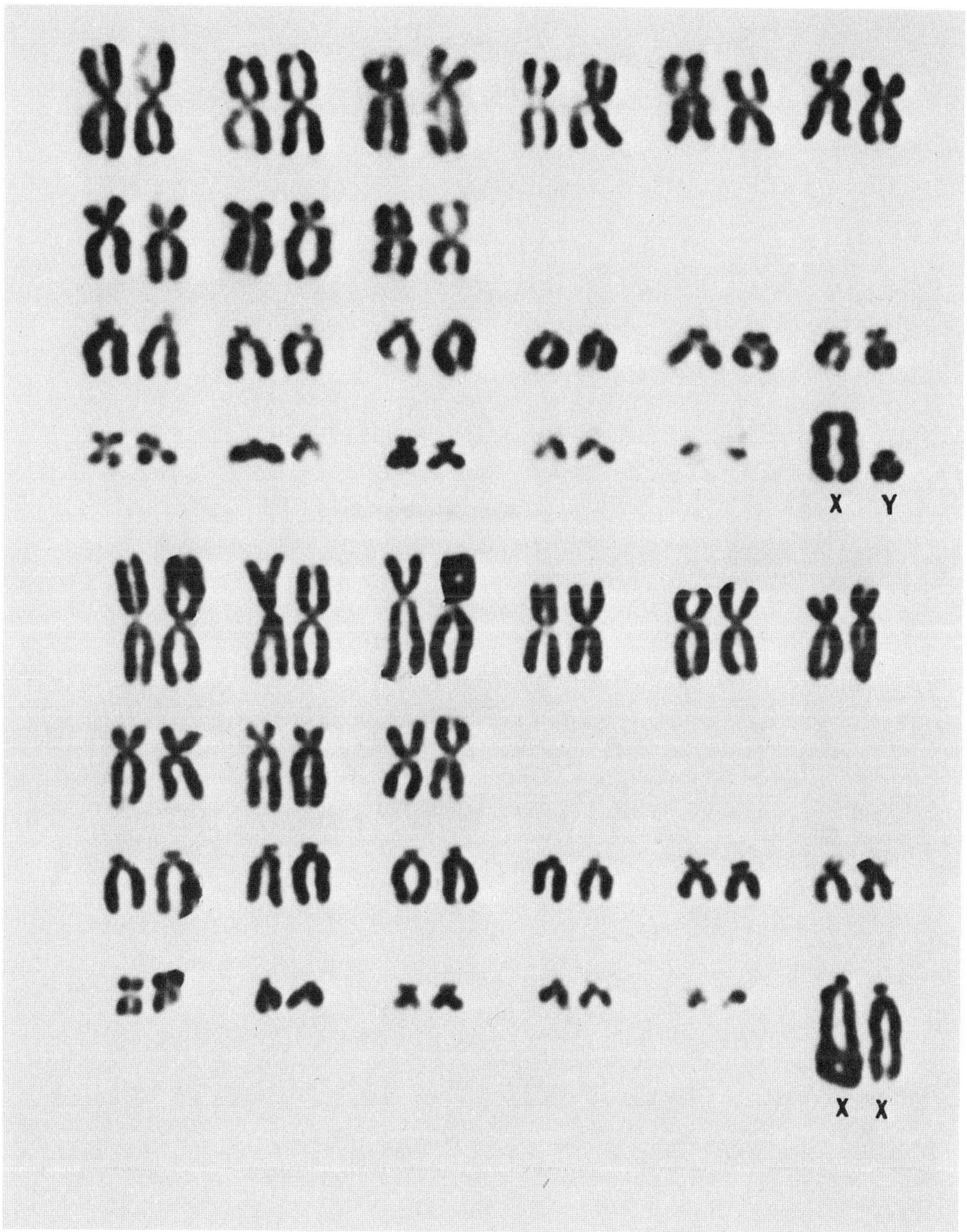

# Order: CHIROPTERA

## Family: PHYLLOSTOMIDAE

***Artibeus turpis*** (Teapa fruit-eating bat)

$2n = 30$

Volume 5, Folio 206, 1971

Order: CHIROPTERA Family: PHYLLOSTOMIDAE

*Artibeus turpis* (Teapa fruit-eating bat)

2n=30

AUTOSOMES: 28 Metacentrics and submetacentrics

SEX CHROMOSOMES: X Submetacentric
Y Metacentric

This is one species in the genus *Artibeus* where a classic XX/XY sex determination system exists. Thus far the karyotypes of all *Artibeus* species (from Central America to the Caribbean Islands) are indistinguishable save for the Y chromosome constitution.

The specimens were collected by Dr. Robert J. Baker in Sinaloa, Mexico. Bone marrows were used for cytological preparations.

REFERENCES:

1) Baker, R.J.: Karyotypes of bats of the family Phyllostomidae and their taxonomic implications. Southwest. Natural. 12:407, 1967.

2) Hsu, T.C., Baker, R.J. and Utakoji, T.: The multiple sex chromosome system of American leaf-nosed bats (Chiroptera, Phyllostomidae). Cytogenetics 7:27-38, 1968.

Artibeus turpis (Teapa fruit-eating bat)

2n=30

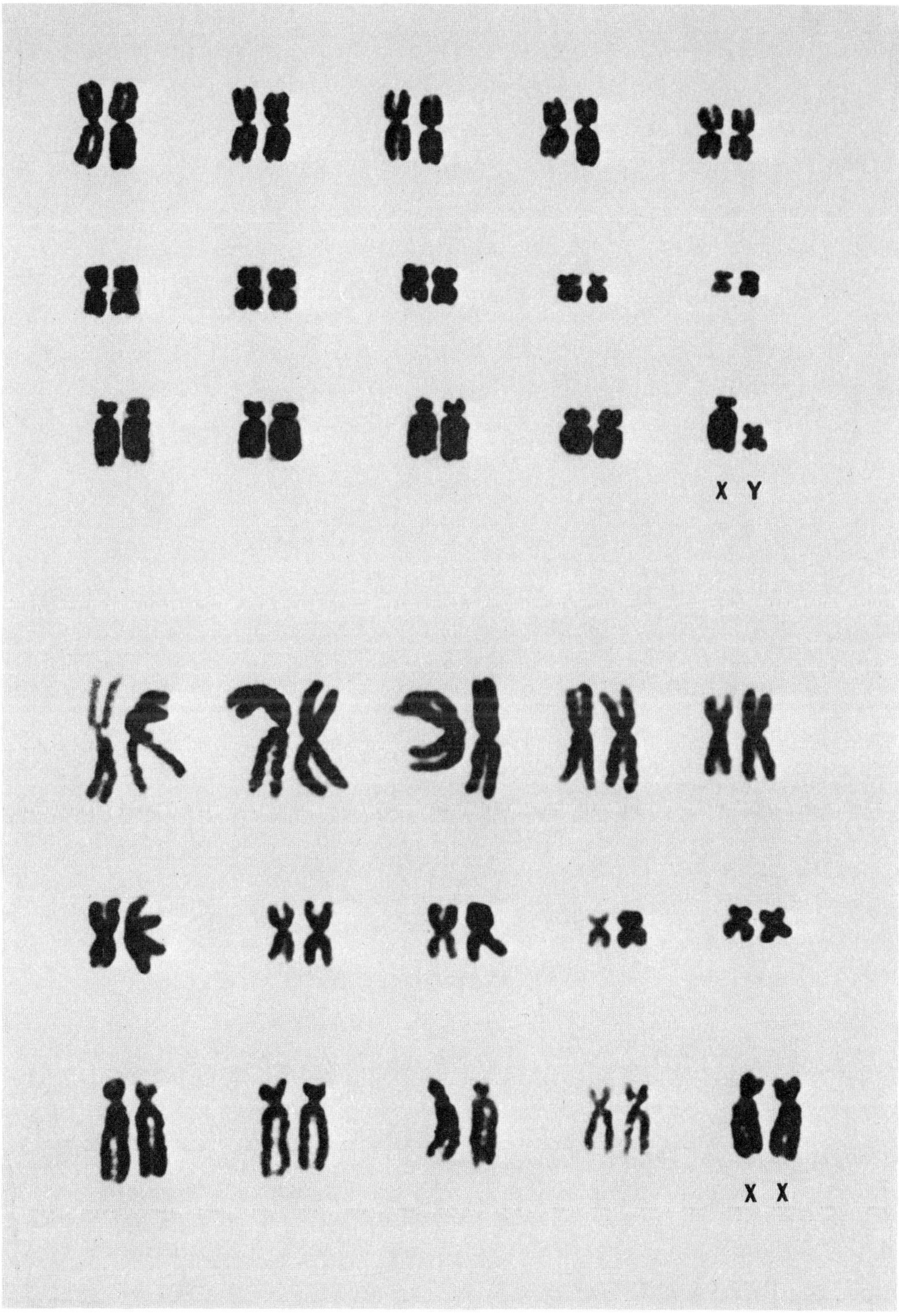

# Order: CHIROPTERA

## Family: PHYLLOSTOMIDAE

### *Chiroderma villosum*

$2n = 26$

Order: CHIROPTERA Family: PHYLLOSTOMIDAE

## *Chiroderma villosum*

2n=26

AUTOSOMES: 24 Metacentrics and submetacentrics

SEX CHROMOSOMES: X Chromosome submetacentric
Y Chromosome submetacentric

The X chromosomes and three pairs of autosomes have distinctly unequal arm ratios. Identification of the X is equivocal. The Y is the smallest element of the complement.

The specimens were collected by Dr. Robert J. Baker from San Fernando Ranch, Chiapas, Mexico. Bone marrows were used for cytological preparations. The specimen number for the male is 34-109172, and the female, 34-109193. Both are in the collection of the Texas Technological University, Lubbock, Texas, USA.

REFERENCES:

1) Baker, R.J.: Karyotypes of bats of the family Phyllostomidae and their taxonomic implications. Southwest. Natural. 12:407, 1967.

2) Hsu, T.C., Baker, R.J. and Utakoji, T.: The multiple sex chromosome system of American leaf-nosed bats (Chiroptera, Phyllostomidae). Cytogenetics 7:27-38, 1968.

# *Chiroderma villosum*

2n=26

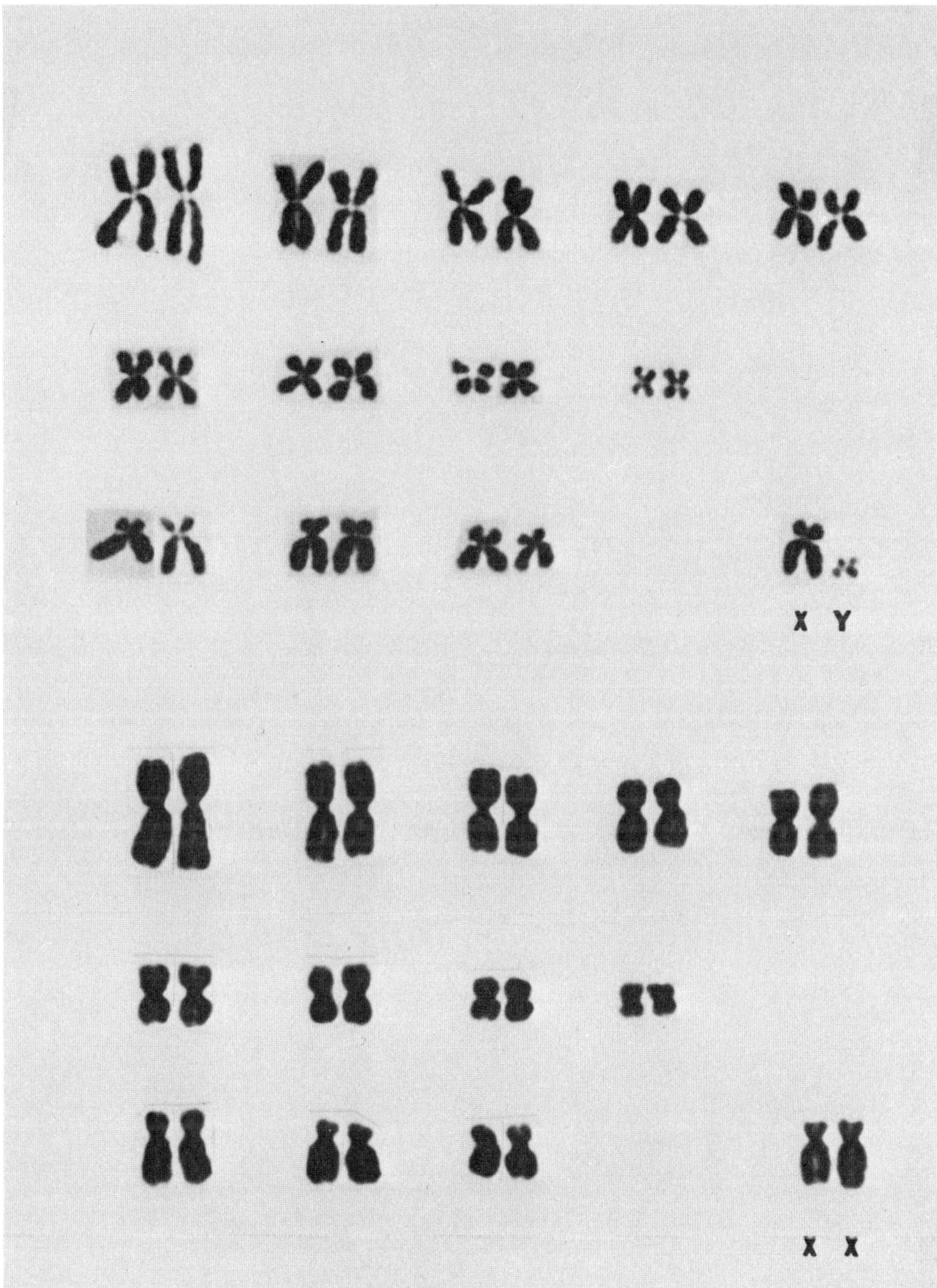

# Order: CHIROPTERA

## Family: PHYLLOSTOMIDAE

### *Glossophaga soricina* (Pallas' long-tongued bat)
### 2n = 32

Order: CHIROPTERA Family: PHYLLOSTOMIDAE

## *Glossophaga soricina* (Pallas' long-tongued bat)

2n=32

AUTOSOMES: 30 Metacentrics and submetacentrics

SEX CHROMOSOMES: X Submetacentric
Y Minute acrocentric

According to Baker (1967), all the species in the genus *Glossophaga* studied thus far, *G. soricina* (Pallas), *G. alticola* Davis, and *G. commissarisi* Gardner have indistinguishable karyotypes. The smallest autosomal pair bears a secondary constriction on one arm near the centromere. Identification of the X is equivocal, but identification of the Y is absolute.

The karyotypes presented here came from specimens collected by Dr. Robert J. Baker. The male specimen was caught 231 Km from Mexico City by Highway 95 to Acapulco, Guerrero, Mexico, and the female, 157 Km by Highway 95 south of Mexico City, Guerrero, Mexico. Bone marrows were used for cytological preparations.

REFERENCES:

1) Baker, R.J.: Karyotypes of bats of the family Phyllostomidae and their taxonomic implications. Southwestern Natural. 12:407, 1967.

2) Hsu, T.C., Baker, R.J. and Utakoji, T.: The multiple sex chromosome system of American leaf-nosed bats (Chiroptera, Phyllostomidae). Cytogenetics 7:27, 1968.

<u>Glossophaga</u> <u>soricina</u> (Pallas' long-tongued bat)

2n=32

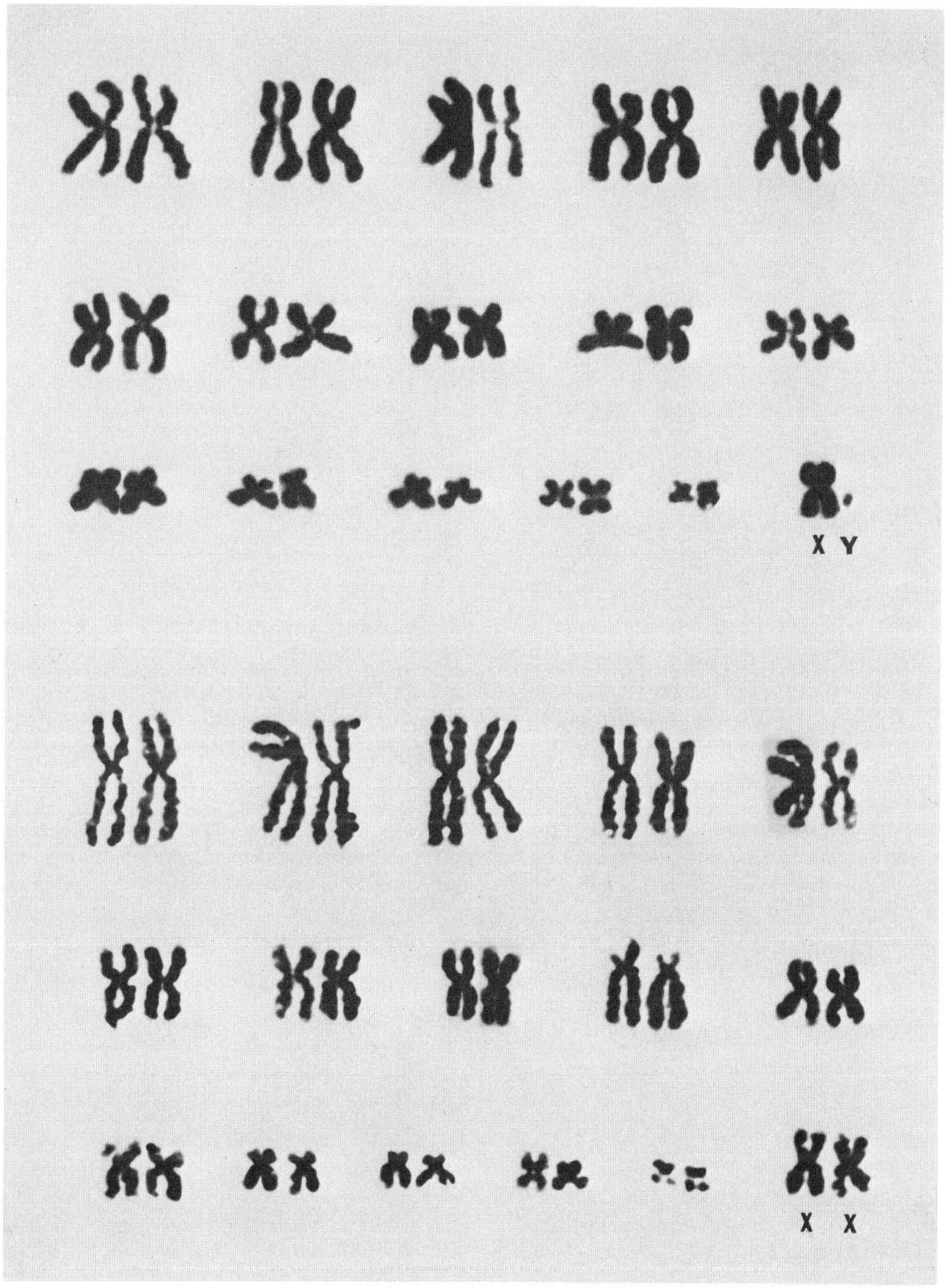

# Order: CHIROPTERA

## Family: PHYLLOSTOMIDAE

### *Pteronotus parnellii*
### 2n = 38

Order: CHIROPTERA Family: PHYLLOSTOMIDAE

# *Pteronotus parnellii*

2n=38

AUTOSOMES: 24 Metacentrics and submetacentrics
12 Acrocentrics

SEX CHROMOSOMES: X Submetacentric
Y Acrocentric

According to Baker (1967), three species of this genus, *Pt. parnellii* (Gray), *Pt. psilotus* (Dobson), and *Pt. davyi* (Gray), possess identical karyotypes. The identification of the X chromosome is somewhat subjective, but it is one of the medium-sized submetacentrics with nearly equal arms. The Y chromosome knob-like second arm of the Y chromosome may be obscure in some cells.

The specimens were collected by Dr. Robert J. Baker. The male was caught 231 Km from Mexico City by Highway 95 to Acapulco, Guerrero, Mexico, and the female, from Km 184 on Highway 200 N. of Huixulta, Chiapas, Mexico. Bone marrows were used for cytological preparations.

REFERENCES:

1) Baker, R.J.: Karyotypes of bats of the family Phyllostomidae and their taxonomic implications. Southwestern Natural. 12:407, 1967.

2) Hsu, T.C., Baker, R.J. and Utakoji, T.: The multiple sex chromosome system of American leaf-nosed bats (Chiroptera, Phyllostomidae). Cytogenetics 7:27, 1968.

Pteronotus parnellii

2n=38

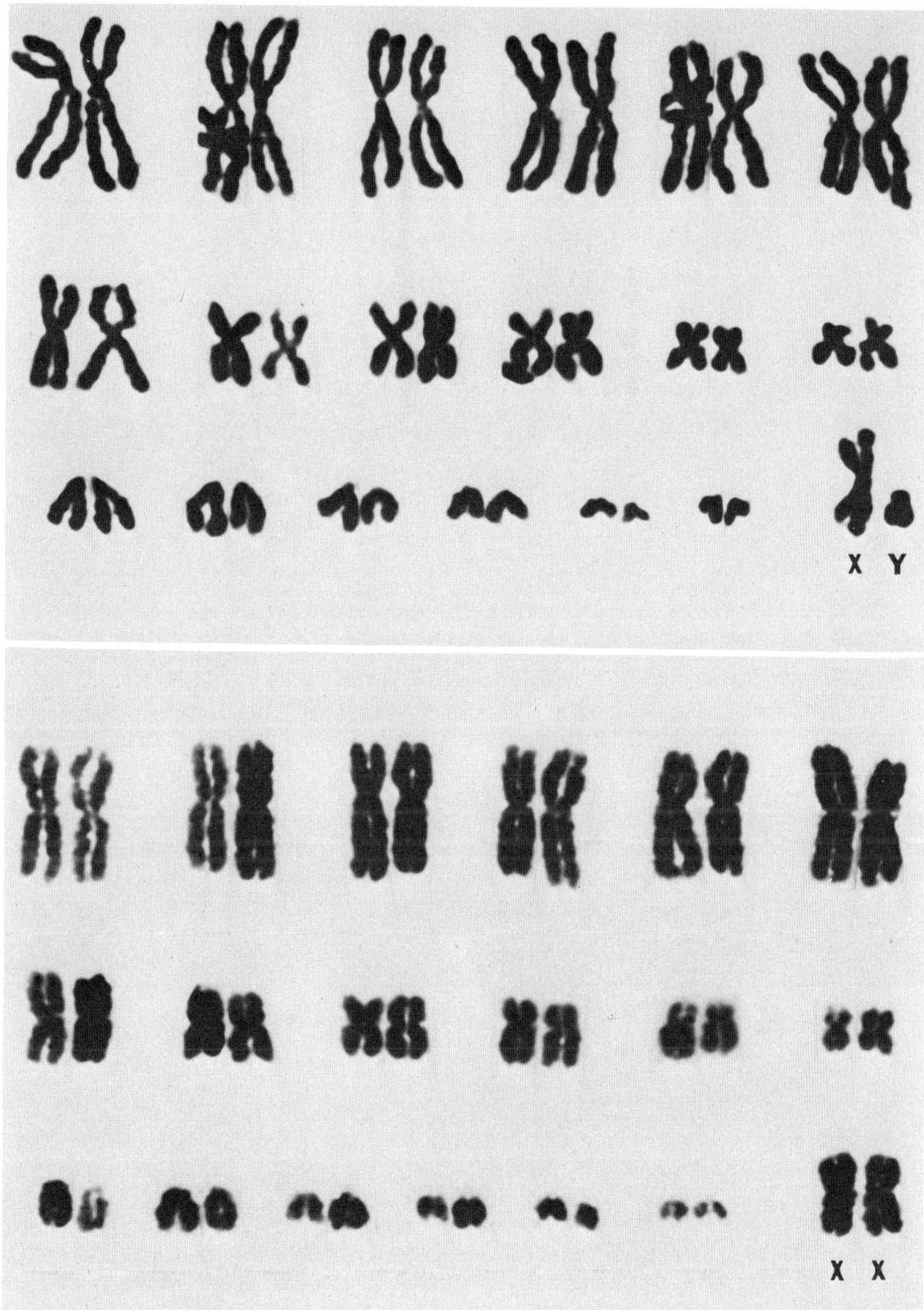

# Order: CHIROPTERA

## Family: VESPERTILIONIDAE

***Nycticeius humeralis*** (Evening bat)
$2n = 46$

Order: CHIROPTERA Family: VESPERTILIONIDAE

*Nycticeius humeralis* (Evening bat)

2n=46

AUTOSOMES: 4 Metacentrics
40 Acrocentrics

SEX CHROMOSOMES: X Submetacentric
Y Minute

One pair of the medium-sized acrocentrics has a secondary constriction near the centromere. The two pairs of large, metacentric autosomes are indistinguishable, but they both are considerably longer than the X. Identification of the X chromosome is therefore unequivocal. Although several pairs of autosomes are very small, the Y is the smallest element of the complement.

The specimens were collected in the vicinity of Houston, Texas, USA. Lung cultures were used for cytological preparations.

REFERENCES:

1) Baker, R.J. and Patton, J.L.: Karyotypes and karyotypic variation of North American Vespertilionid bats. J. Mammal. 48:270, 1967.

Order: CHIROPTERA Family: VESPERTILIONIDAE

# *Nycticeius humeralis* (Evening bat)

2n=46

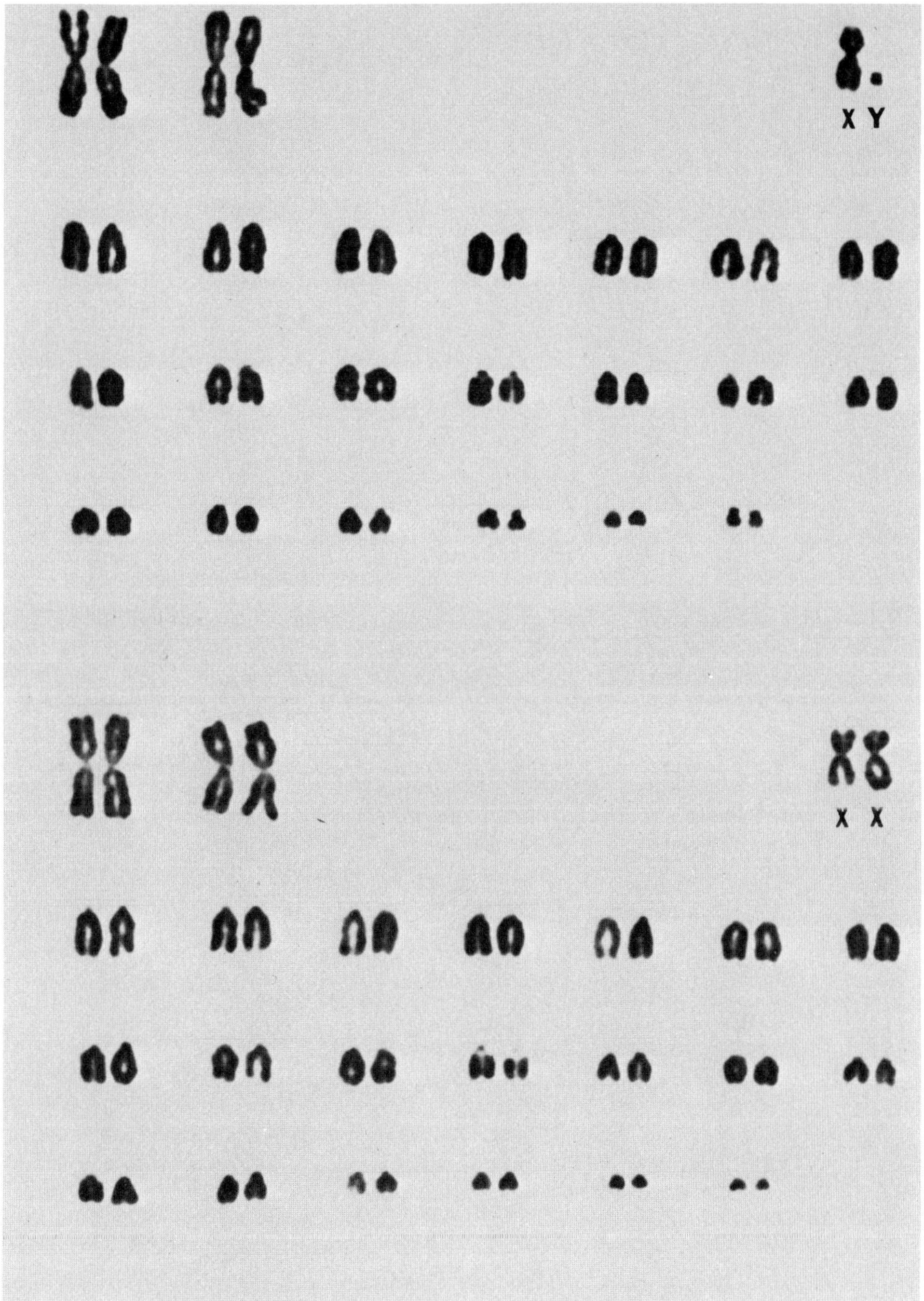

# Order: LAGOMORPHA

## Family: LEPORIDAE

*Lepus townsendii* (White-tailed jack rabbit)
$2n = 48$

Order: LAGOMORPHA Family: LEPORIDAE

## *Lepus townsendii* (White-tailed jack rabbit)

2n=48

AUTOSOMES: 42 Metacentrics and submetacentrics
4 Acrocentrics

SEX CHROMOSOMES: X Submetacentric
Y Acrocentric

Skin biopsies of two animals trapped at Ferguson County, Montana, USA, were kindly made available by Dr. B. W. O'Gara, Missoula, Montana. As with the two previously displayed jack rabbits (Folios 6, 7) the short arms of many autosomes differ in size and classification as to acrocentrics or subtelocentrics is highly arbitrary. These three species have very similar, if not identical, karyotypes.

REFERENCES:

1) Jalal, S.M., James, T.R. and Seabloom, R.W.: Karyotype of white-tailed jack rabbit. Mammalian Chromosomes Newsletter 8:194, 1967.

Order: LAGOMORPHA Family: LEPORIDAE

<u>Lepus townsendii</u> (White-tailed jack rabbit)

2n=48

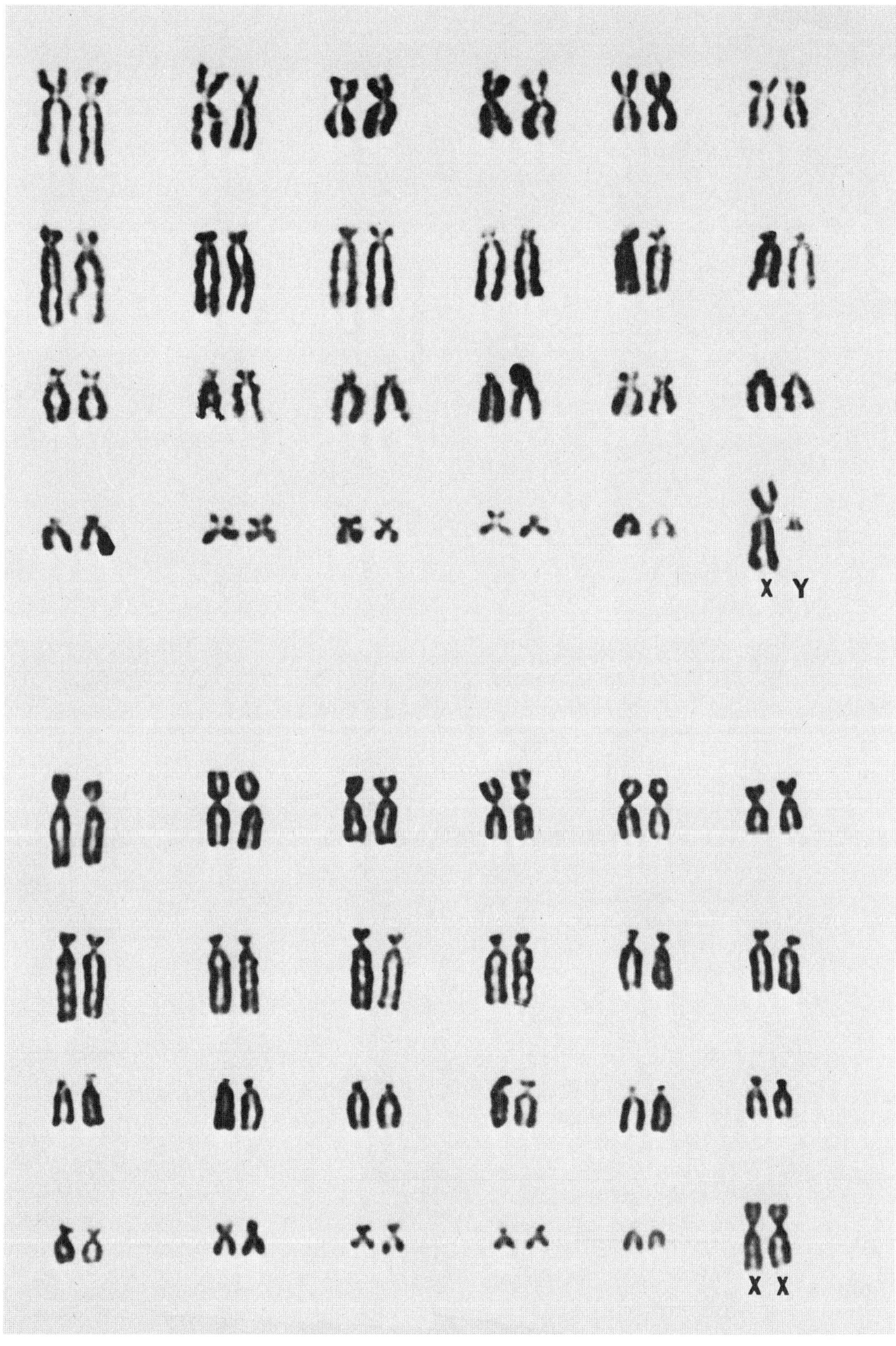

# Order: LAGOMORPHA

## Family: LEPORIDAE

***Sylvilagus floridanus*** (Eastern cottontail)
2n = 42

Order: LAGOMORPHA Family: LEPORIDAE

*Sylvilagus floridanus* (Eastern cottontail)

2n=42

AUTOSOMES: 40 Metacentrics, submetacentrics and subtelocentrics

SEX CHROMOSOMES: X Submetacentric
Y Acrocentric

At least four pairs of subtelocentrics bear satellites on the short arm. Identification of the X chromosome is ambiguous by morphology alone, but it is one of the medium-sized submetacentrics. The Y is the smallest element of the complement.

The male specimen was obtained in Houston, Texas, USA. Lung cultures were initiated for cytological preparations. The female karyotype presented here is a gift of Dr. Katherine G. Palmer.

REFERENCES:

1) Palmer, C. and Armstrong, R.: Chromosome number and karyotype of *Sylvilagus floridanus*, the Eastern cottontail. Mammalian Chromosomes Newsletter 8:282, 1967.

2) Holden, H.E.: The karyotypes of two species of *Sylvilagus*. Mammalian Chromosomes Newsletter 9:230, 1968.

*Sylvilagus floridanus* (Eastern cottontail)

2n=42

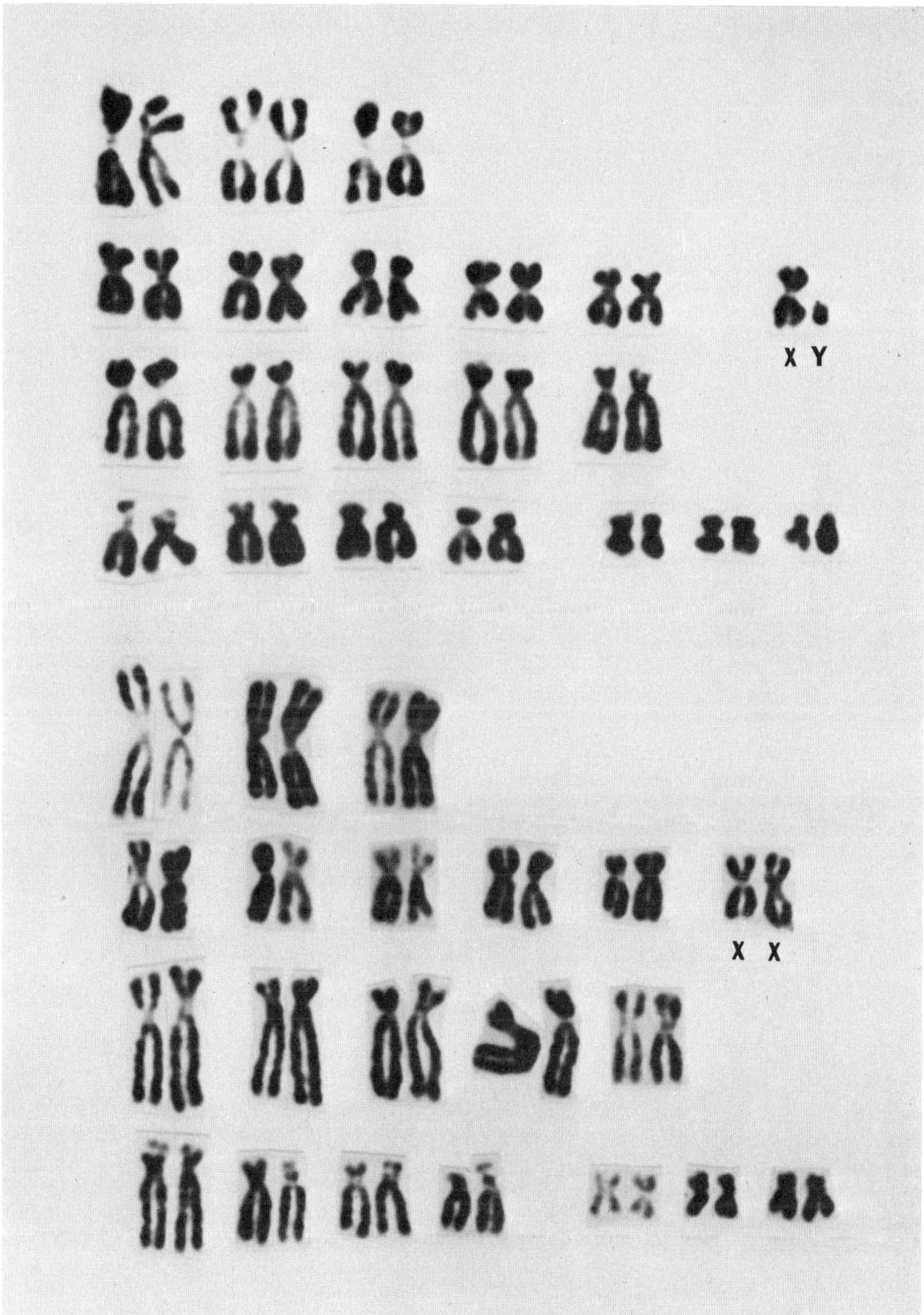

# Order: RODENTIA

## Family: SCIURIDAE

***Ammospermophilus harrisii*** (Harris' antelope squirrel)
$2n = 32$

Order: RODENTIA Family: SCIURIDAE

*Ammospermophilus harrisii* (Harris' antelope squirrel)

2n=32

AUTOSOMES: 30 Metacentrics and submetacentrics

SEX CHROMOSOMES: X Submetacentric
Y Acrocentric

The X chromosome is one of the large submetacentric elements, but its identification is not unequivocal. The Y chromosome is the only acrocentric of the complement.

The specimens were collected 1 mile north of Kingman, Mohave County, Arizona, USA. Bone marrows were used for cytological preparations. The specimens are stored in Texas Technological University, Lubbock, Texas, USA, bearing specimen numbers ♀ TCH 1248, ♂ TCH 1249.

REFERENCES:

1) Nadler, C.F. and Sutton, D.A.: Mitotic chromosomes of some North American Sciuridae. Proc. Soc. Exp. Biol. & Med. 110:36, 1962.

Ammospermophilus harrisii (Harris' antelope squirrel)

2n=32

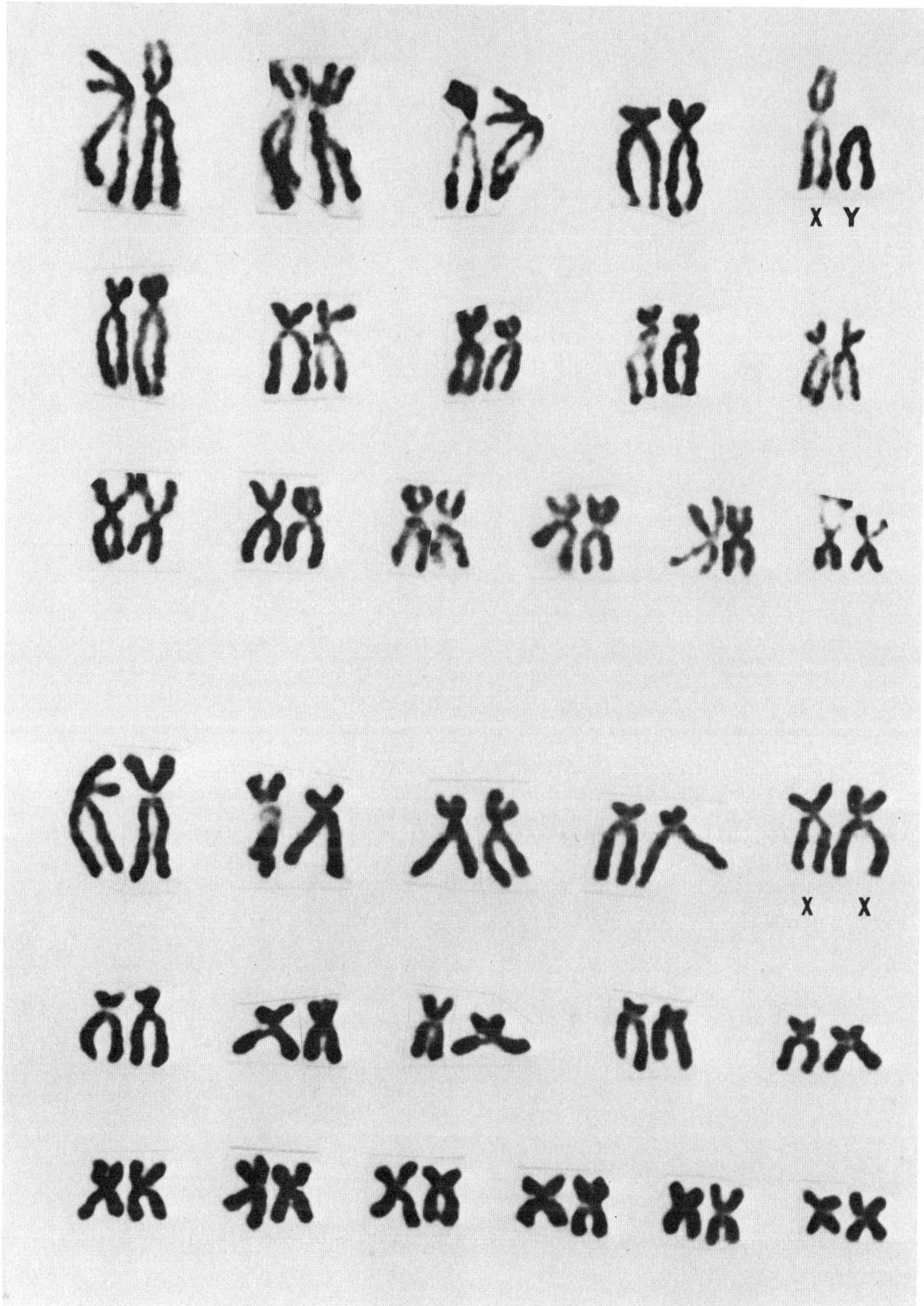

# Order: RODENTIA

## Family: SCIURIDAE

***Callospermophilus (Citellus) lateralis***

(Golden-mantled ground squirrel)

2n = 42

Order: RODENTIA Family: SCIURIDAE

*Callospermophilus* (*Citellus*) *lateralis*
(Golden-mantled ground squirrel)

2n=42

AUTOSOMES: 40 Metacentrics and submetacentrics

SEX CHROMOSOMES: X Submetacentric
Y Acrocentric

Skin biopsies from two animals trapped in Missoula County, Montana, USA, were kindly made available by Dr. B. W. O'Gara.

REFERENCES:

1) Nadler, C.F. and Sutton, D.A.: Mitotic chromosomes of some North American Sciuridae. Proc. Soc. Exp. Biol. Med. 110:36, 1962.

2) Nadler, C.F.: Chromosomes of *Spermophilus franklini* and taxonomy of the ground squirrel genus *Spermophilus*. Syst. Zool. 15:199, 1966.

Callospermophilus (Citellus) lateralis
(Golden-mantled ground squirrel)

2n=42

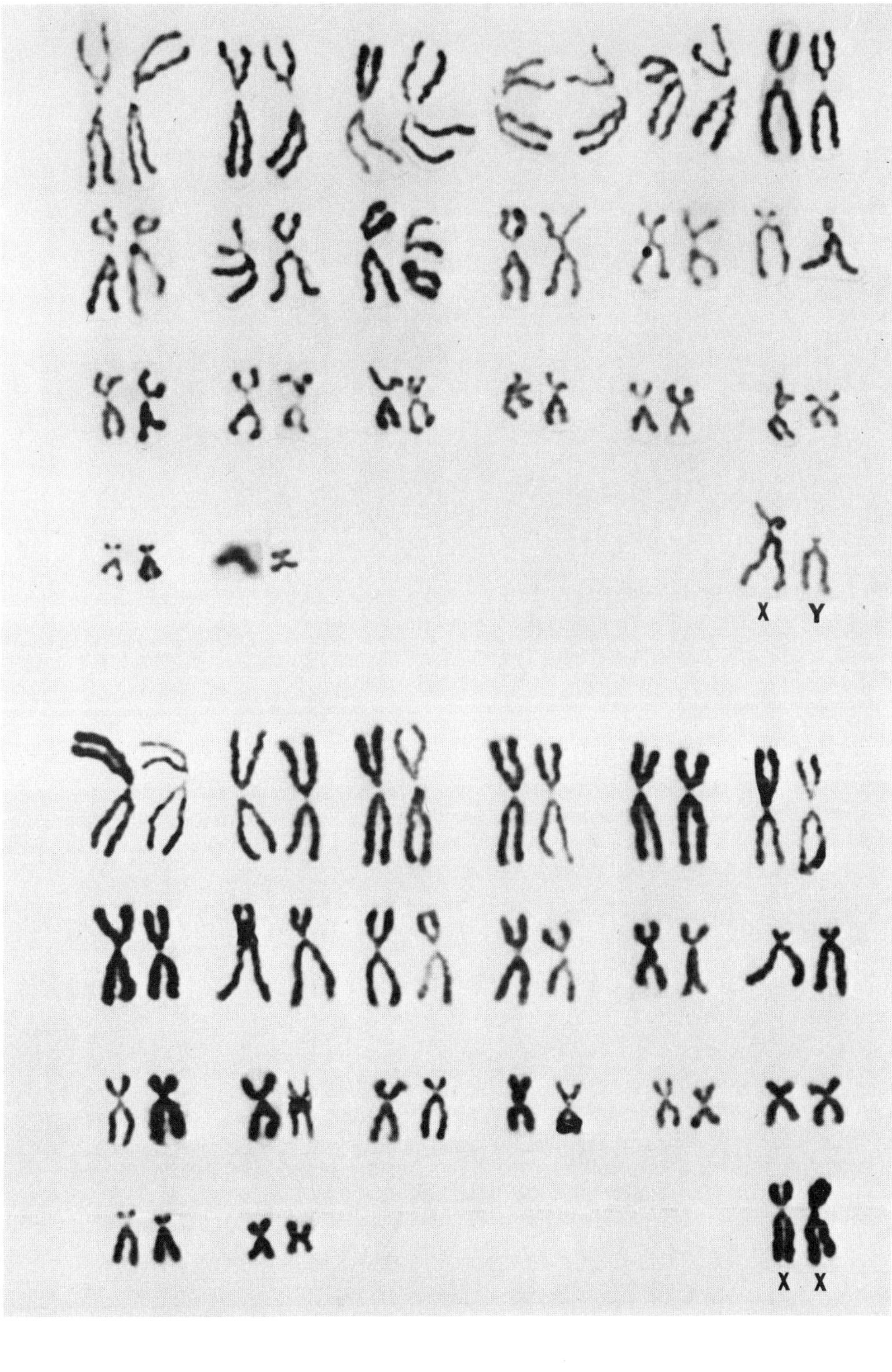

# Order: RODENTIA

## Family: SCIURIDAE

*Cynomys gunnisoni* (Gunnison's Prairie dog)
2n = 40

Order: RODENTIA Family: SCIURIDAE

*Cynomys gunnisoni* (Gunnison's Prairie dog)

2n=40

AUTOSOMES: 26 Metacentrics and submetacentrics
12 Subtelocentrics or acrocentrics

SEX CHROMOSOMES: X Submetacentric
Y Acrocentric

Most of the "subtelocentrics or acrocentrics" have distinct short arms. The X chromosome is among the large submetacentrics, and the Y, the smallest acrocentric.

The karyotypes presented here are gifts of Dr. Charles F. Nadler, Northwestern University Medical School, Chicago, Illinois, USA. The specimens belong to the subspecies *C. g. zuniensis* Hollister, and were collected from Flagstaff, Coconino County, Arizona, USA, by Dr. Terry A. Vaughan. Bone marrows were used for cytological preparations.

REFERENCES:

1) Nadler, C.F. and Hoffmann, R.S.: Chromosomes and serum proteins of some prairie dogs (Cynomys). J. Mammal. (in press)

Cynomys gunnisoni (Gunnison's Prairie dog)

2n=40

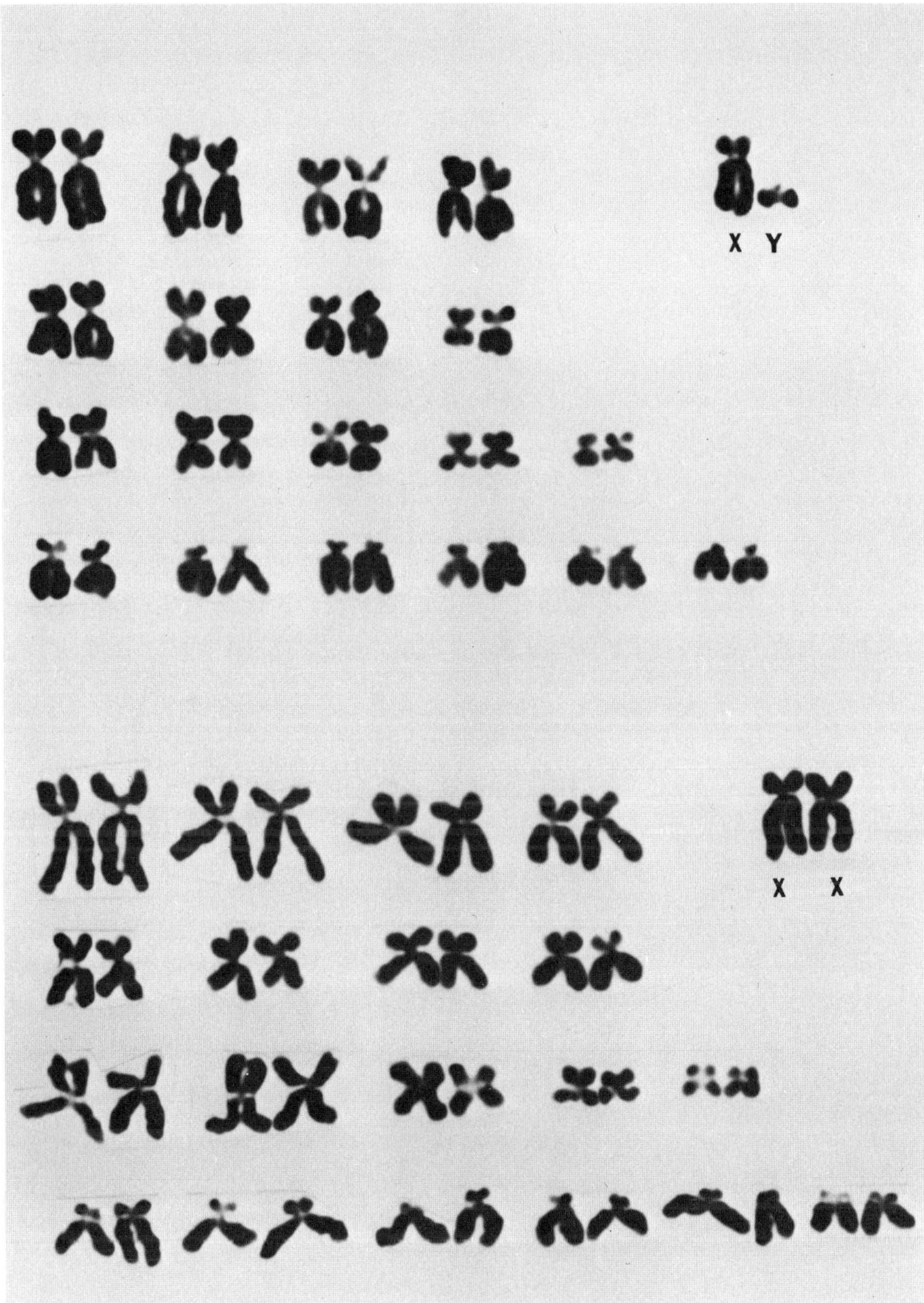

# Order: RODENTIA

## Family: SCIURIDAE

***Eutamias ruficaudus*** (Red-tailed chipmunk)
2n = 38

Order: RODENTIA Family: SCIURIDAE

*Eutamias ruficaudus* (Red-tailed chipmunk)

2n=38

AUTOSOMES: 22 Metacentrics and submetacentrics
14 Acrocentrics

SEX CHROMOSOMES: X Submetacentric
Y Metacentric

Lung, skin and spleen biopsies of two male and two female animals trapped in Missoula County, Montana, USA, were kindly made available by Dr. B. W. O'Gara. The karyotype of the male came from a spleen culture, and that of the female from a skin culture. At least one of the acrocentric elements displayed here has distinct but small short arm. Other investigators may prefer to consider this a subtelocentric or even a submetacentric element.

REFERENCES:

1) Nadler, C. F.: Contributions of chromosomal analysis to the systematics of North American chipmunks. Amer. Midland Nat. 72:298, 1964.

2) Sutton, D. A. and Nadler, C. F.: Chromosomes of the North American chipmunk genus *Eutamias*. J. Mammal. 50:524, 1969.

Eutamias ruficaudus (Red-tailed chipmunk)

2n=38

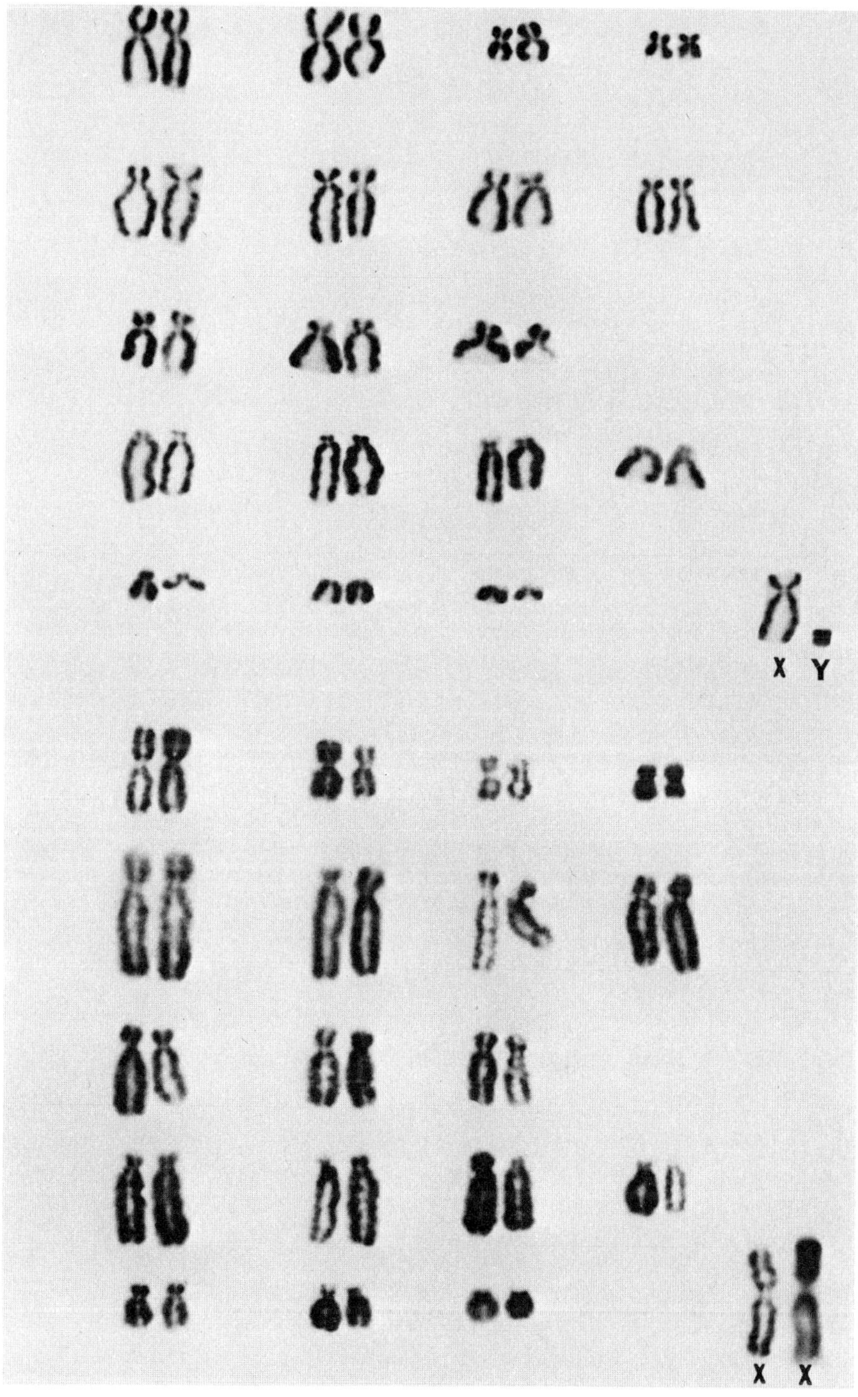

# Order: RODENTIA

## Family: SCIURIDAE

***Eutamias sibiricus*** (Asian chipmunk)
2n = 38

Order: RODENTIA Family: SCIURIDAE

*Eutamias sibiricus* (Asian chipmunk)

2n=38

AUTOSOMES: 20 Metacentrics, submetacentrics and subtelocentrics
16 Acrocentrics

SEX CHROMOSOMES: X Submetacentric
Y Metacentric

Several pairs of autosomes can be identified without great difficulty. The acrocentrics can be roughly classified into two groups: A group of three pairs of larger elements and another of five pairs of smaller elements. The X chromosome, however, resembles some of the autosomes.

The karyotypes are gifts of Dr. Charles F. Nadler. Another set of karyotypes donated by Dr. M. Sasaki showed similar results.

REFERENCES:

1) Nadler, C.F., Hoffmann, R.S. and Lay, D.M.: Chromosomes of the Asian chipmunk *Eutamias sibiricus* Laxmann (Rodentia, Sciuridae). Experientia 25:868, 1969.

2) Sasaki, M., Shimba, H. and Itoh, M.: Karyotypes of two species of Asiatic squirrels. Mammalian Chromosomes Newsletter 10:227, 1969.

*Eutamias sibiricus* (Asian chipmunk)

2n=38

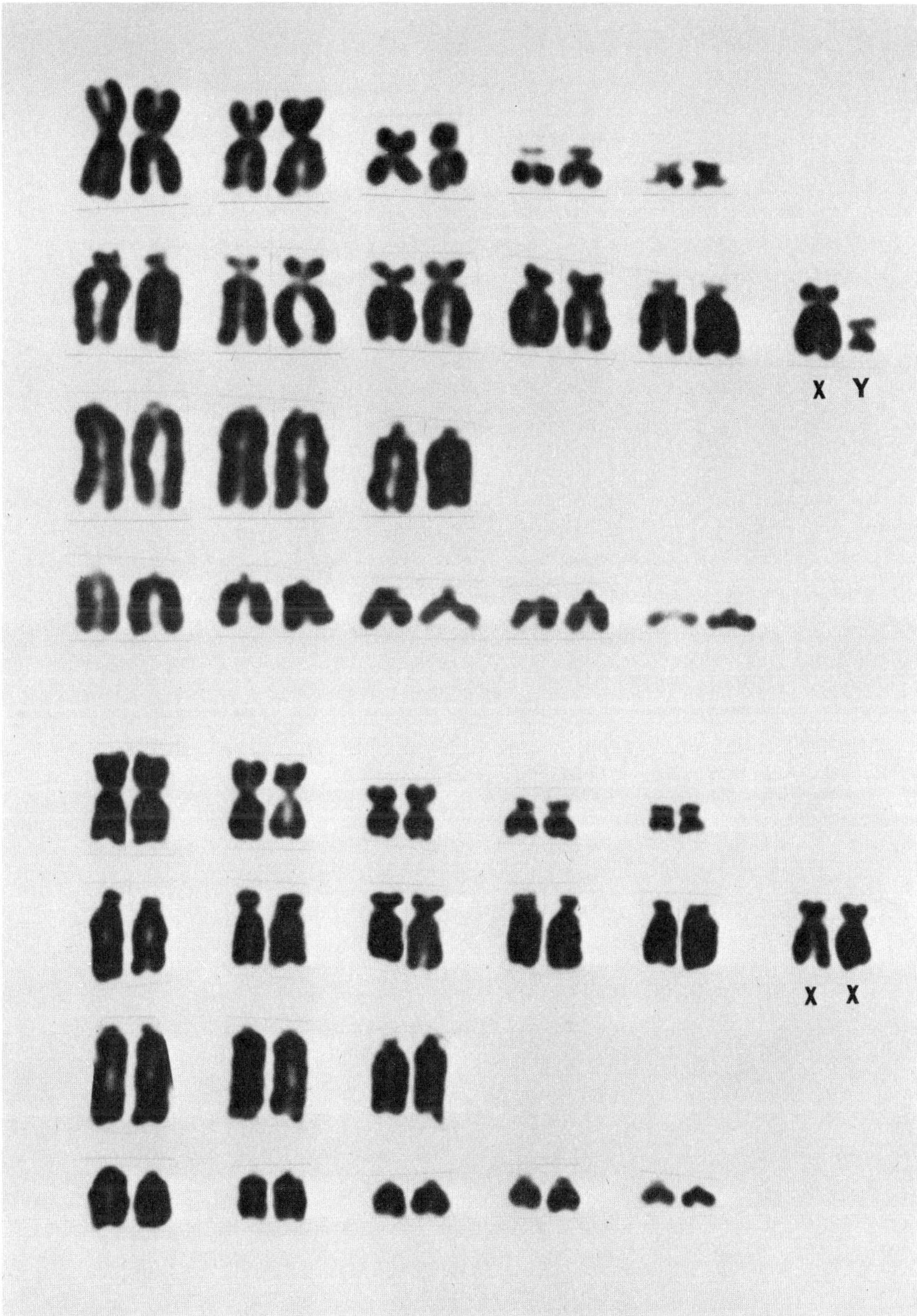

# Order: RODENTIA

## Family: CRICETIDAE

***Peromyscus floridanus*** (Florida mouse)

$2n = 48$

Order: RODENTIA Family: CRICETIDAE

*Peromyscus floridanus* (Florida mouse)

2n=48

AUTOSOMES: 16 Submetacentrics
30 Acrocentrics

SEX CHROMOSOMES: X Submetacentric with highly disproportioned arms
Y Small submetacentric

These specimens were collected in the vicinity of Sumner, Florida, USA, in 1967. Samples from Delray Beach, Florida, USA, collected by Dr. J.A. King showed identical karyotypes. All cytological preparations were made from primary lung cultures.

The B-group chromosomes (see Hsu and Arrighi, 1968), are placed in the last of the karyotypes presented here. The X chromosome appears to be the second longest submetacentric of the complement. Identification of the Y offers no great difficulty.

REFERENCES:

1) Hsu, T.C. and Arrighi, F.E.: Chromosomes of *Peromyscus* (Rodentia, Cricetidae). I. Evolutionary trends in 20 species. Cytogenetics 7:417, 1968.

Peromyscus floridanus (Florida mouse)

2n=48

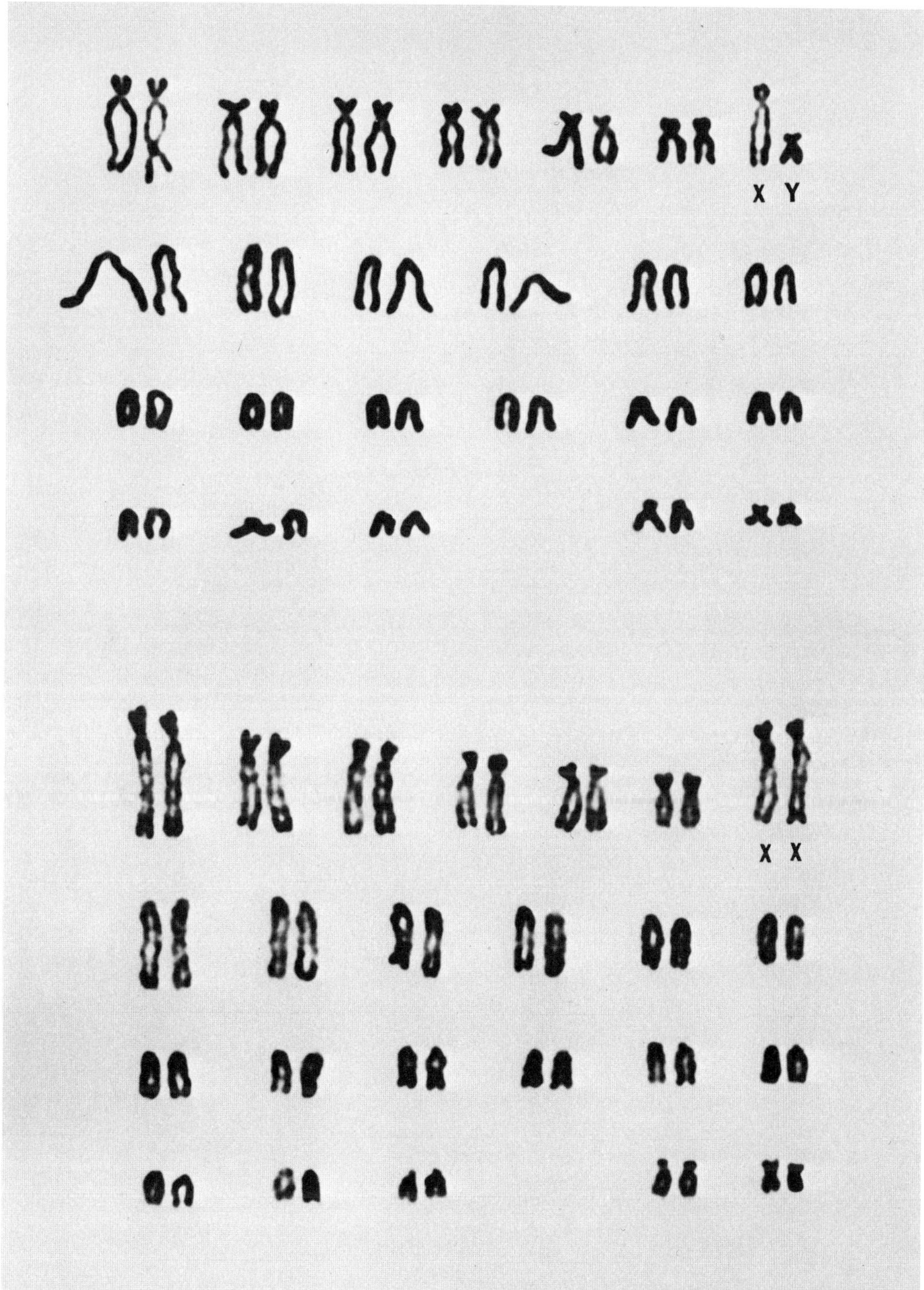

# Order: RODENTIA

## Family: CRICETIDAE

***Peromyscus melanophrys*** (Plateau mouse)
$2n = 48$

Order: RODENTIA Family: CRICETIDAE

*Peromyscus melanophrys* (Plateau mouse)

2n=48

AUTOSOMES: 10 Metacentrics and submetacentrics
36 Acrocentrics

SEX CHROMOSOMES: X Submetacentric
Y Acrocentric

According to the nomenclature system of Hsu and Arrighi, the biarmed autosomes are, from left to right in the front row, $A_1$, $A_2$, $A_3$, $B_1$, $B_2$. From a limited number of specimens analyzed, it is evident that polymorphism in respect to chromosome features exists. The Y chromosome may vary in size, and the number of biarmed autosomes may be higher than 10 in some individuals.

The male specimen was collected by Dr. J. L. Patton in 1966 at Zecatecas, Mexico, and the female was the second generation descendent of animals originally caught by Dr. Rollin H. Baker in 1967 at 5 miles South of Rio Grande, Murchison Ranch, Mexico. Lung cultures were initiated for cytological preparations.

REFERENCES:

1) Hsu, T.C. and Arrighi, F.E.: Chromosomes of *Peromyscus* (Rodentia, Cricetidae). I. Evolutionary trends in 20 species. Cytogenetics 7:417, 1968.

Peromyscus melanophrys (Plateau mouse)

2n=48

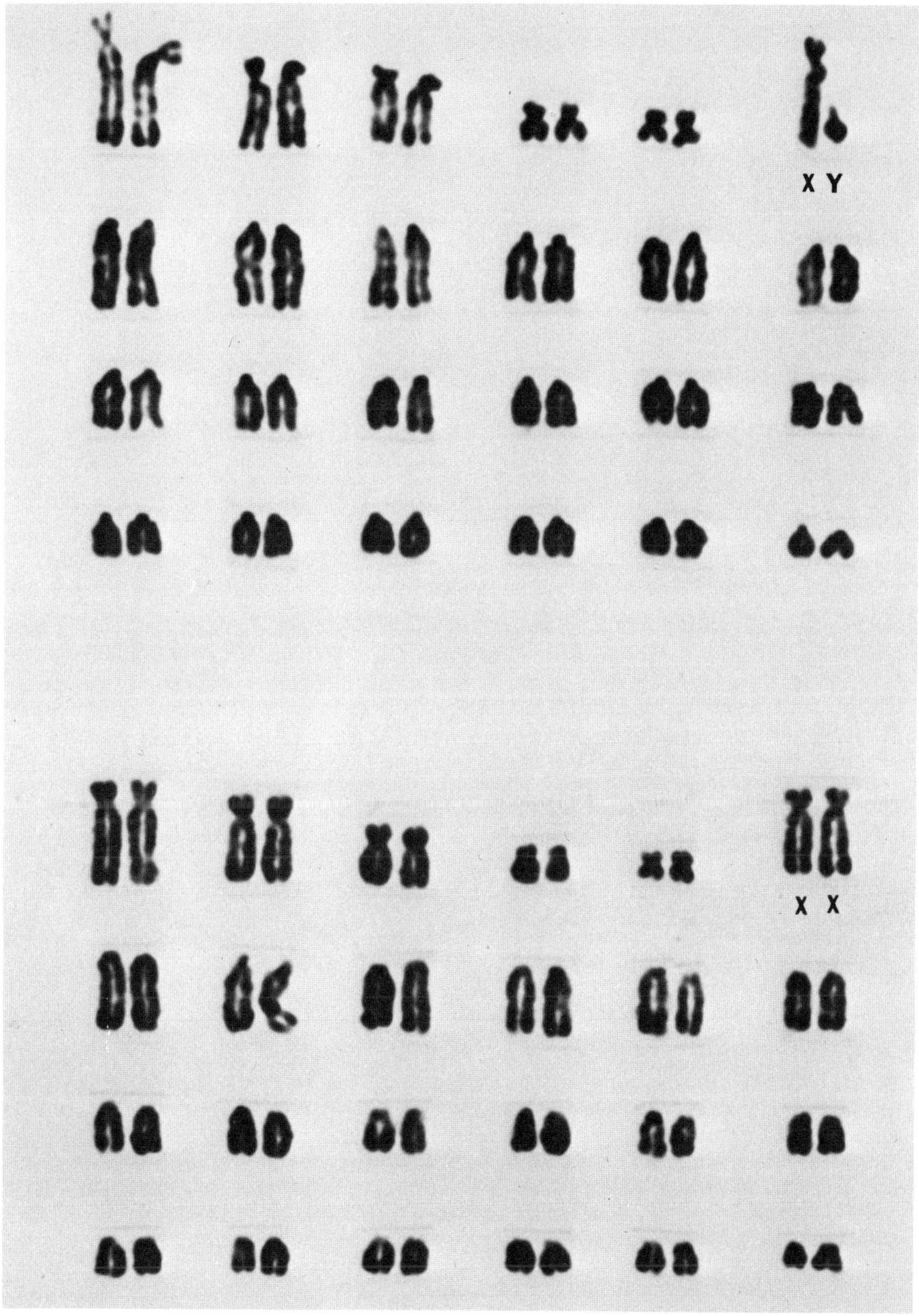

# Order: RODENTIA

## Family: CRICETIDAE

*Arvicola terrestris* (Water vole)

2n = 36

Order: RODENTIA Family: CRICETIDAE

Arvicola terrestris (Water vole)

2n=36

AUTOSOMES: 26 Metacentrics and submetacentrics
8 Acrocentrics

SEX CHROMOSOMES: X Submetacentric
Y Acrocentric

These karyotypes were kindly donated by Dr. K. Fredga, Lund, Sweden, and came from tissue cultures of various organs of five male and two female specimens from different parts of Sweden. All were similar. Four chromosomes may display secondary constrictions and satellites, and satellite association occurs in acrocentrics. A trisomic animal has been described by Fredga. The karyotype arrangement follows his description.

REFERENCES:

1) Matthey, R.: Analyse cytotaxonomique de huit espèces de Muridés. Murinae, Cricetinae, Microtinae paléarctiques et nord-américains. Arch. Klaus-Stift. Vererb.-Forsch. 32:385, 1957.

2) Fredga, K.: Idiogram and trisomy of the water vole (Arvicola terrestris L.), a favorable animal for cytogenetic research. Chromosoma 25: 75, 1968.

*Arvicola terrestris* (Water vole)

2n=36

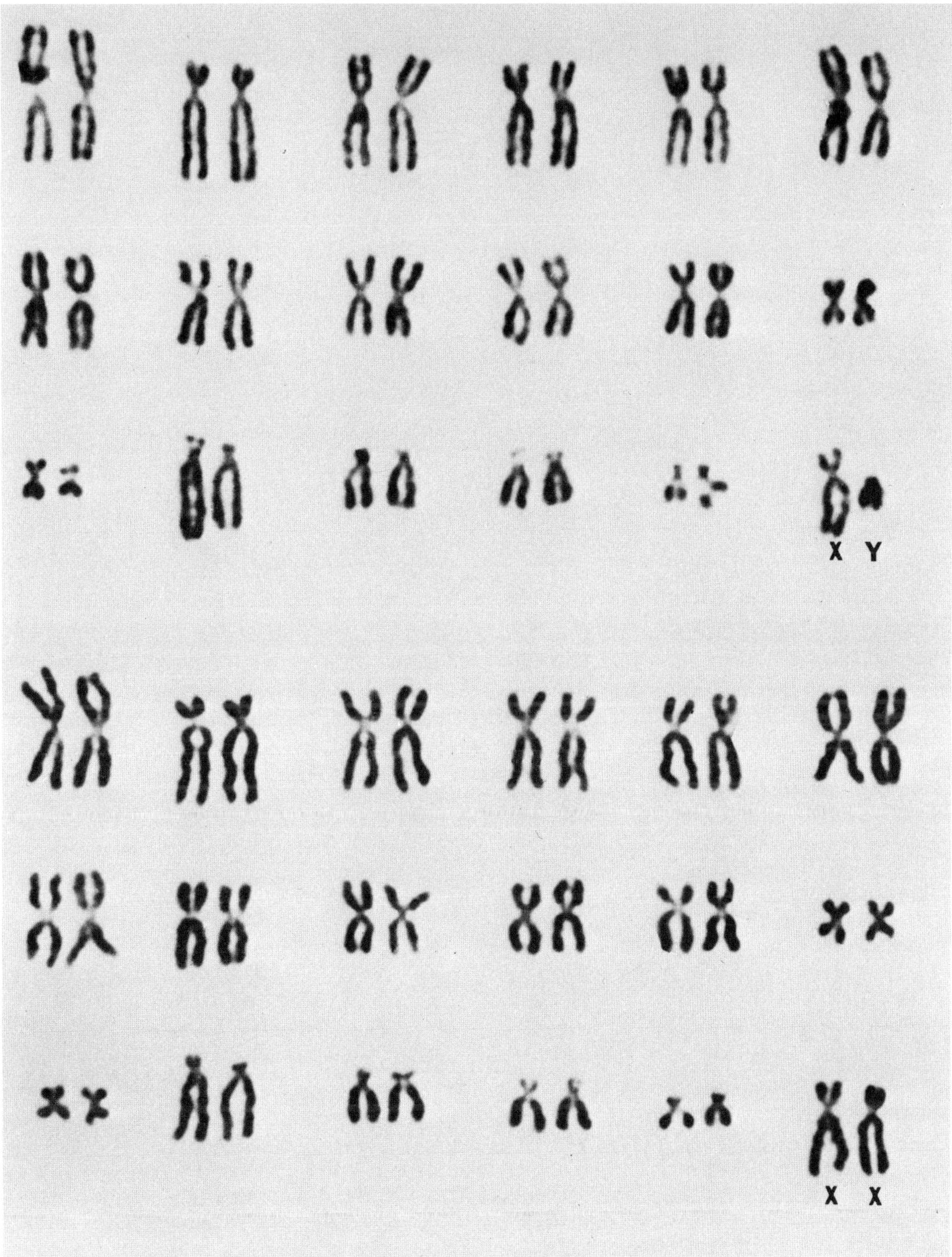

# Order: RODENTIA

## Family: CRICETIDAE

***Lagurus curtatus*** (Sagebrush vole)
$2n = 54$

*Lagurus curtatus* (Sagebrush vole)

2n=54

AUTOSOMES: 2 Submetacentrics
50 Acrocentrics

SEX CHROMOSOMES: X Submetacentric
Y Acrocentric

Identification of the X chromosomes is unequivocal, but identification of the Y is subjective. Among the acrocentric autosomes, only one pair, the longest, can be separated from the rest.

The specimens were kindly donated by Dr. Murray L. Johnson. The male was collected in Jefferson County, Oregon, USA, and the female, from East Mansfield, Douglas County, Washington, USA. Lung cultures were initiated for cytological preparations.

<u>Lagurus</u> <u>curtatus</u> (Sagebrush vole)

2n=54

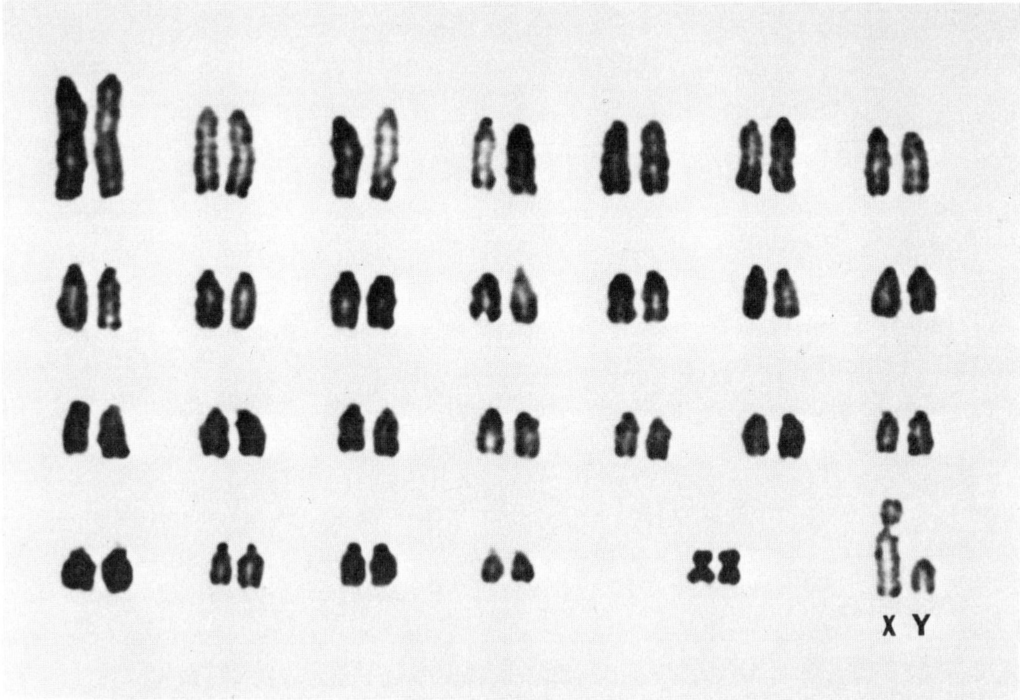


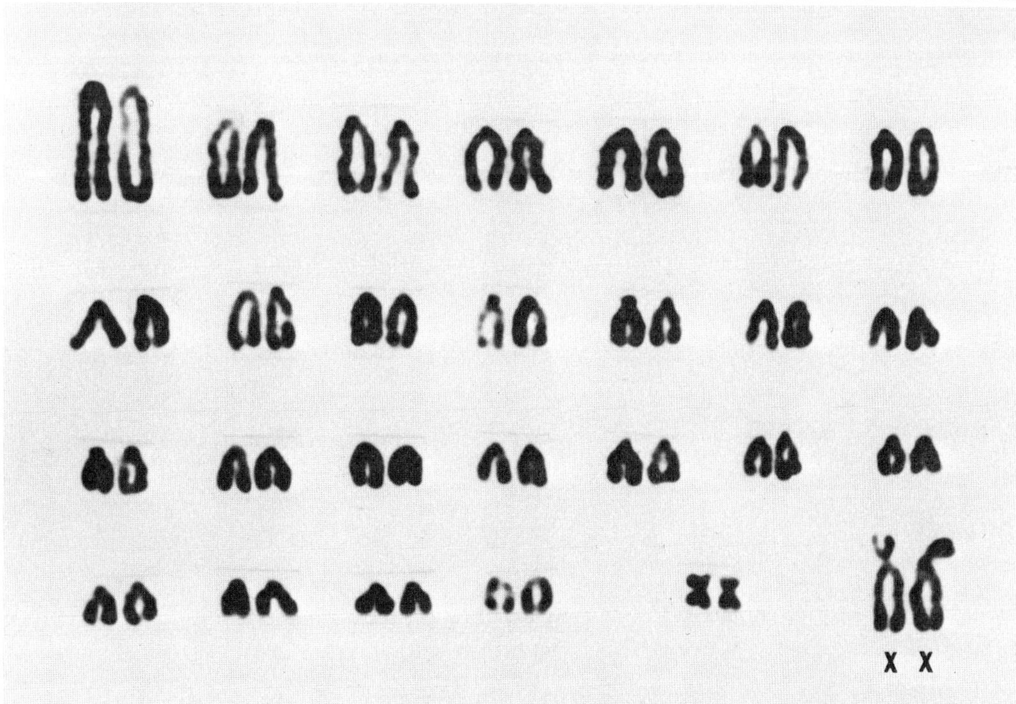

# Order: RODENTIA

## Family: CRICETIDAE

***Lagurus lagurus*** (Steppe lemming)
$2n = 54$

Order: RODENTIA                                                    Family: CRICETIDAE

*Lagurus lagurus* (Steppe lemming)

2n=54

AUTOSOMES: 8 Metacentrics and submetacentrics
44 Acrocentrics

SEX CHROMOSOMES: X Submetacentric
Y Acrocentric

All the four pairs of biarmed chromosomes are morphologically distinguishable. The identification of the X chromosome is unequivocal; but the identification of the Y is not. The Y is apparently one of the smallest acrocentrics.

The karyotypes presented here are gifts of Dr. K. Kernahle, Department of Biology, Martin-Luther-Universität, DDR402 Halle/S., Universitätsplatz 7., Germany. The animals were descendents of four pairs imported in 1963 from Moscow, USSR.

REFERENCES:


1) Matthey, R.: Analyse cytotaxonomique de huit espèces de Muridés. *Murinae Cricetinae*, *Microtinae* paléarctiques et nord-américains. Arch. Julius Klaus-Stiftung 32:26, 1957.

2) Matthey, R.: Les chromosomes des Mammifères euthériens. Arch. Julius-Klaus-Stiftung 33:253, 1958.

3) Pogosianz, H.E. und A.F. Zaharov: On the karyotype of the steppe lemming (*Lagurus lagurus* Pallas). Z. Versuchstierkd. 1:93, 1962.

4) Zernahle, K.: Zur Zytogenetik des Steppenlemmings (*Lagurus lagurus* Pallas). Der Kerngeschlechtdimorphismus und die Chromosomen in der Mitose. In Vorbereitung, 1969.

Lagurus lagurus (Steppe lemming)

2n=54

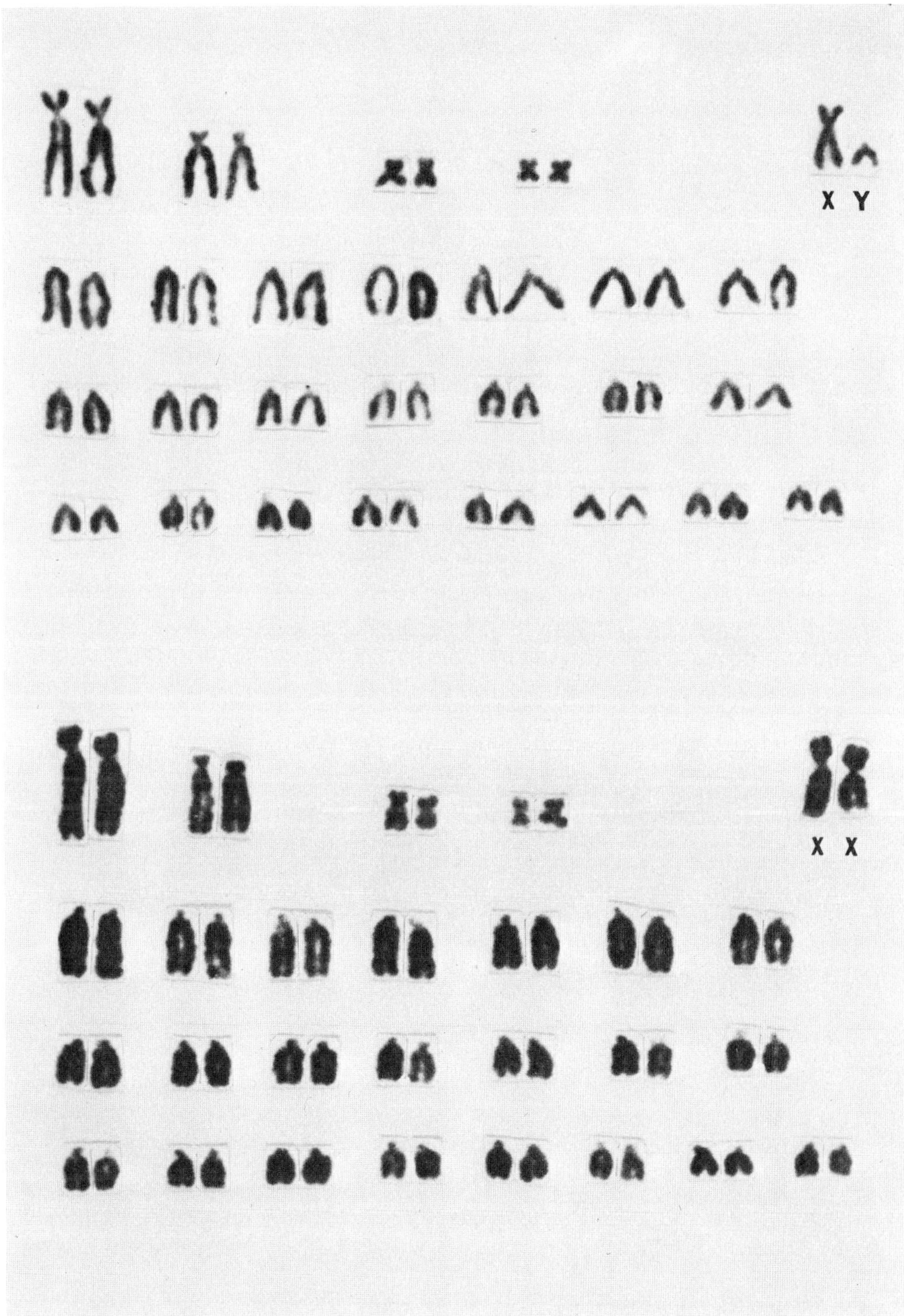

# Order: RODENTIA

## Family: CRICETIDAE

***Microtus townsendii*** (Townsend's vole)
$2n = 50$

Order: RODENTIA Family: CRICETIDAE

*Microtus townsendii* (Townsend's vole)

2n=50

AUTOSOMES: 48 Acrocentrics

SEX CHROMOSOMES: X Acrocentric
Y Acrocentric

The entire complement forms a smooth gradation of sizes without a significant break, so that pairing chromosomes is extremely subjective. The X is apparently one of the larger elements, and the Y, one of the smaller elements.

The specimens (2♀♀, 2♂♂) were kindly donated by Dr. J. Mary Taylor, University of British Columbia, Vancouver, B.C., Canada. They were trapped in the campus of the University of British Columbia. Lung cultures were initiated for cytological preparations. Two additional specimens, 1♀, 1♂, from Pierce County, Washington, USA, supplied by Dr. Murray L. Johnson, gave similar results.

Order: RODENTIA

Family: CRICETIDAE

Microtus townsendii (Townsend's vole)

2n=50

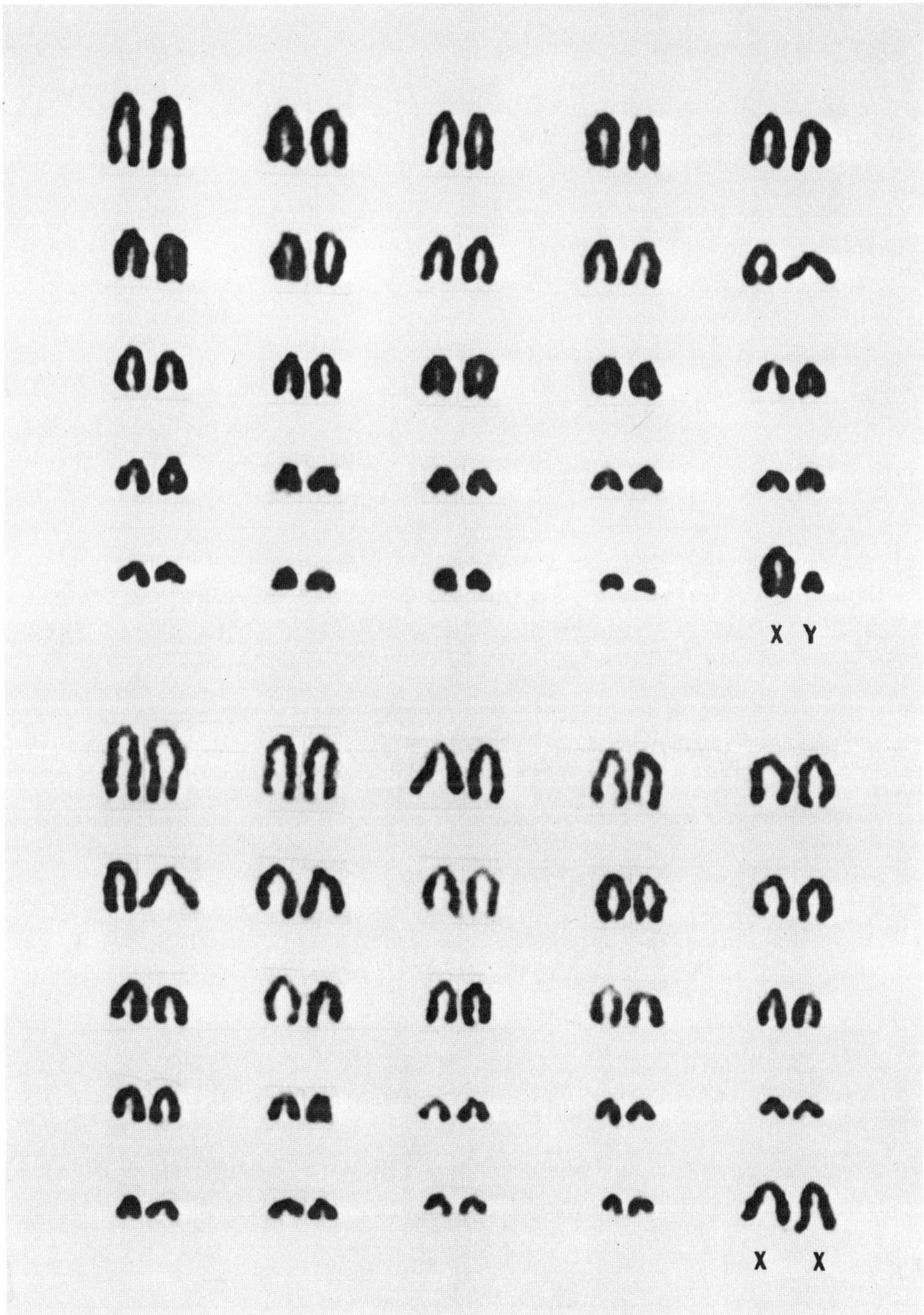

# Order: RODENTIA

## Family: CRICETIDAE

### *Ondatra zibethica* (Muskrat)

$2n = 54$

Order: RODENTIA Family: CRICETIDAE

# *Ondatra zibethica* (Muskrat)

2n=54

AUTOSOMES: 50 Acrocentrics
2 Small submetacentrics

SEX CHROMOSOMES: X Acrocentric
Y Acrocentric

Among the acrocentric autosomes at least one pair is satellited. This is clearly demonstrated in the male karyotype here. The X chromosomes are the longest acrocentrics. This conclusion is ascertained not only from morphology, but also from tritiated thymidine autoradiography.

The karyotypes presented here are gifts of Professor Alfred Gropp. The animals were trapped near the mouth of the Sieg River in the Rhine, approximately 10 miles from Bonn, Germany.

REFERENCES:


1) Matthey, R.: Nouvelles recherches sur les chromosomes des Muridae. Caryologia 6:1, 1954.

2) Moore, W., Jr., Elder, R.L. and Gillespie, L.J.: The chromosomes of the muskrat. J. Hered. 57:104, 1966.

3) Gropp, A. and Geisler, M.: Chromosomes of the muskrat (*Ondatra zibethica* L., Muridae, Rodentia). Mammalian Chromosomes Newsletter 8:286, 1967.

4) Sutton, D.A.: Chromosomes of the muskrat (*Ondatra zibethica* L., Microtinae, Rodentia). Mammalian Chromosomes Newsletter 9:244, 1969.

5) Sutton, D.A.: Chromosomes of a male muskrat (*Ondatra zibethica* L., Microtinae, Rodentia). Mammalian Chromosomes Newsletter 11:24, 1970.

6) Dewald, G.W. and Jalal, S.M.: Karyotypes of two subspecies of North American muskrat. Mammalian Chromosomes Newsletter 11:37, 1970.

Ondatra zibethica (Muskrat)

2n=54

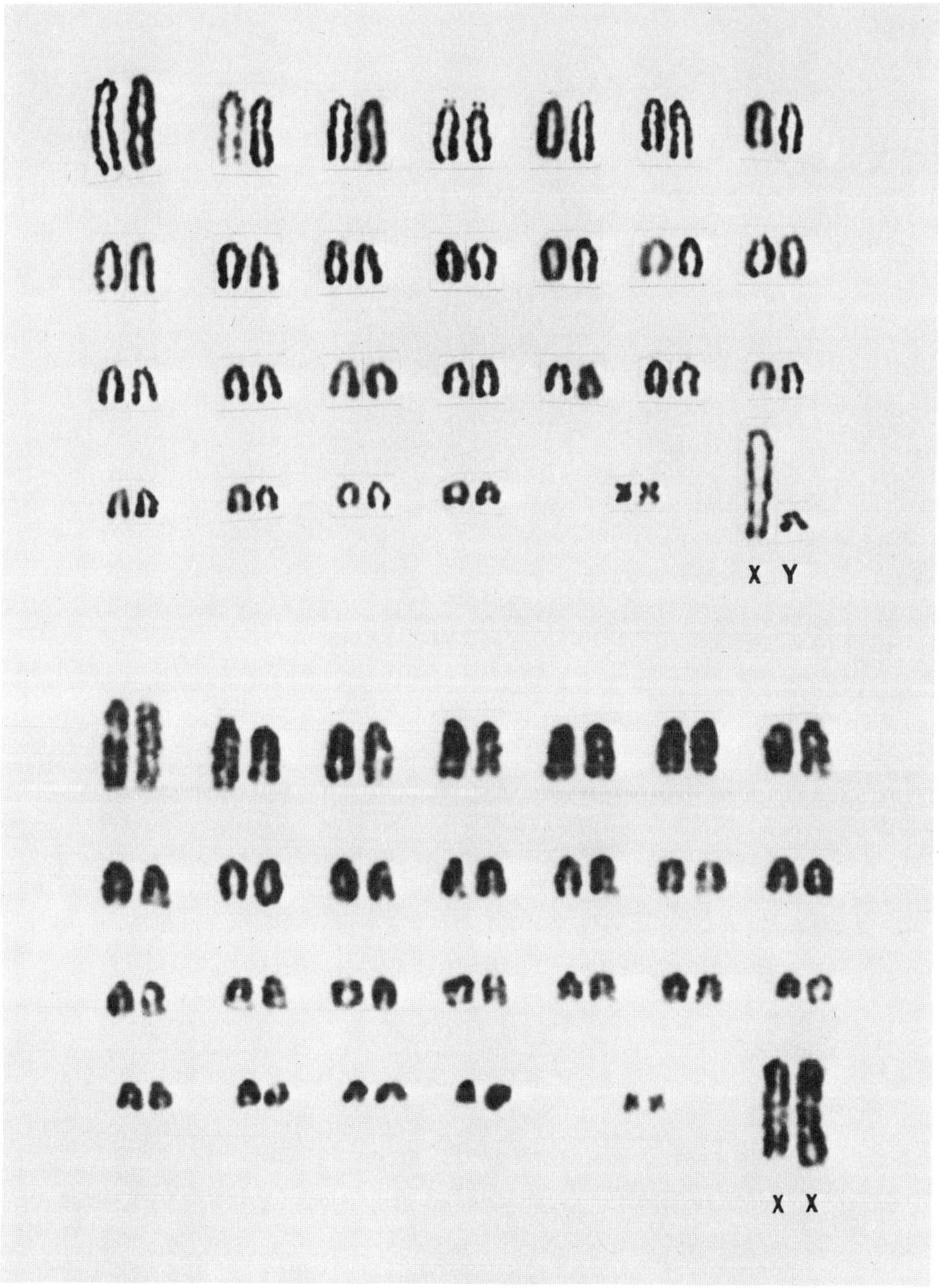

# Order: RODENTIA

## Family: CRICETIDAE

***Meriones hurrianae*** (Indian desert gerbil)

2n = 40

Order: RODENTIA Family: CRICETIDAE

*Meriones hurrianae* (Indian desert gerbil)

2n=40

AUTOSOMES: 36 Metacentrics and submetacentrics
2 Acrocentrics

SEX CHROMOSOMES: X Submetacentric
Y Submetacentric

Many autosome pairs are morphologically similar and they form a more or less continuous size gradation. The X is one of the longest chromosomes in the complement.

The karyotypes presented here are gifts of Dr. Charles F. Nadler. The animals were caught in Gizri (2 miles E. Karachi), Pakistan. Bone marrows were used for cytological preparations.

REFERENCES:

1) Nadler, C.F. and Lay, D.M.: Chromosomes of some species of *Meriones* (Mammalia: Rodentia). Z. f. Säugetierkunde 32:285, 1967.

Meriones hurrianae (Indian desert gerbil)

2n=40

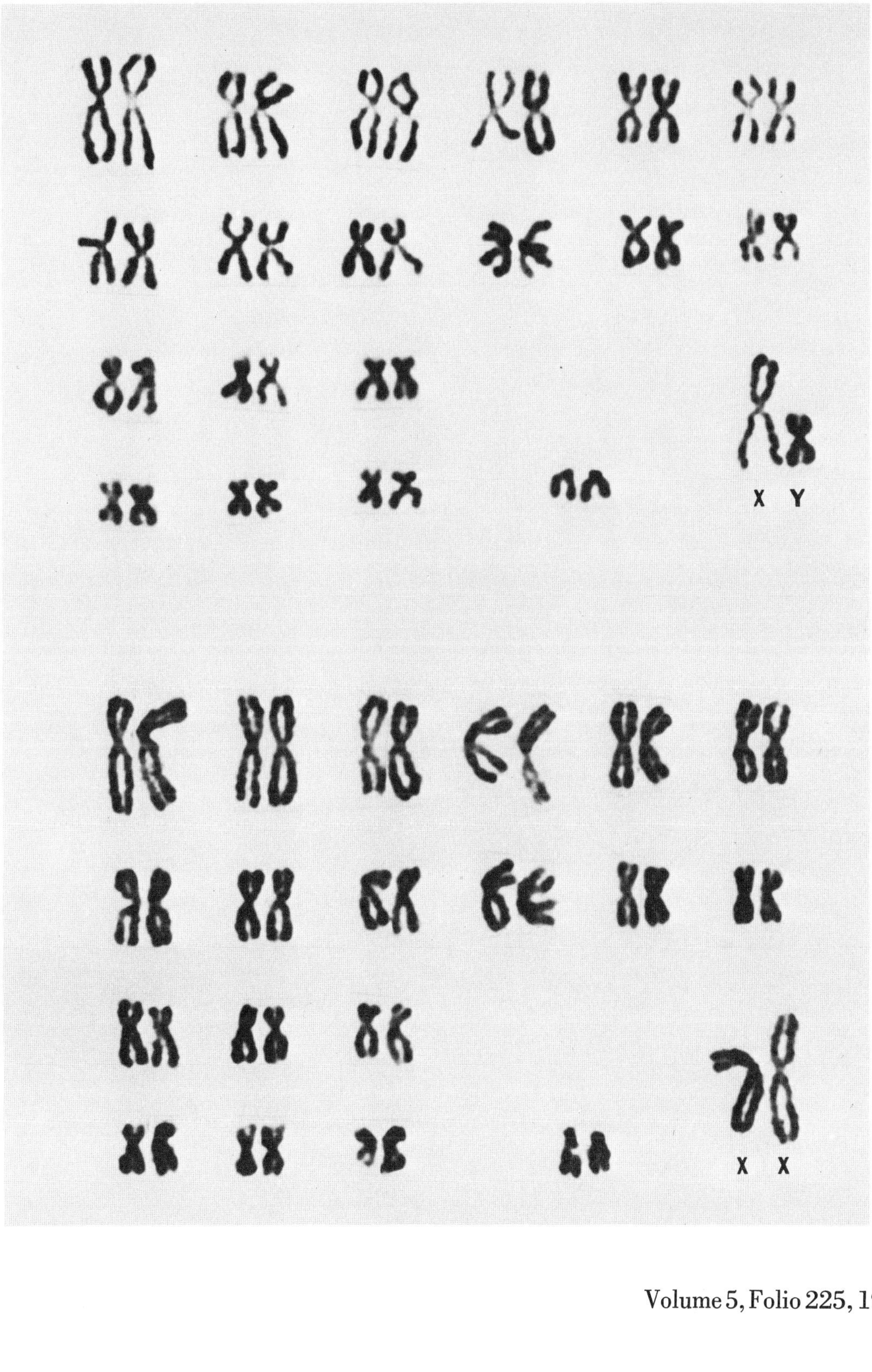

# Order: RODENTIA

## Family: CRICETIDAE

***Meriones libycus*** (Libyan jird)
$2n = 44$

Order: RODENTIA Family: CRICETIDAE

# *Meriones libycus* (Libyan jird)

2n=44

AUTOSOMES: 30 Metacentrics and submetacentrics
12 Acrocentrics

SEX CHROMOSOMES: X Acrocentric
Y Submetacentric

The long acrocentric X chromosomes are characteristic. It can be used as an excellent marker in cytological studies in hybrids between *M. shawi* and *M. libycus*. The two species also differ in several autosomal characteristics, though the diploid numbers are the same. For *M. shawi*, cf. Folio 227.

The karyotypes presented here are gifts of Dr. Charles F. Nadler. The specimens were from laboratory stock originally obtained from Lichtenstein-Rachtagan, Iran.

REFERENCES:

1) Matthey, R.: Les Chromosomes des Muridae. Rev. Suisse de Zool. 60: 225, 1953.

2) Matthey, R.: Cytologie et taxonomie du genre *Meriones*, Illiger. Säugetierk. Mitt. 5:145, 1957.

3) Nadler, C.F. and Lay, D.M.: Chromosomes of some species of *Meriones* (Mammalia: Rodentia). Z. f. Säugetiekunde 32:285, 1967.

4) Lay, D.M. and Nadler, C.F.: Hybridization in the rodent genus *Meriones*. I. Breeding and cytological analyses of *Meriones shawi* (♀) X *Meriones libycus* (♂) hybrids. Cytogenetics 8:35, 1969.

5) Nadler, C.F.: Chromosomal Evolution in Rodents. In "Comparative Mammalian Cytogenetics" (Benirschke, K., ed.), pp. 277-309, Springer-Verlag, New York, 1969.

Order: RODENTIA

Family: CRICETIDAE

<u>Meriones</u> <u>libycus</u> (Libyan jird)

2n=44

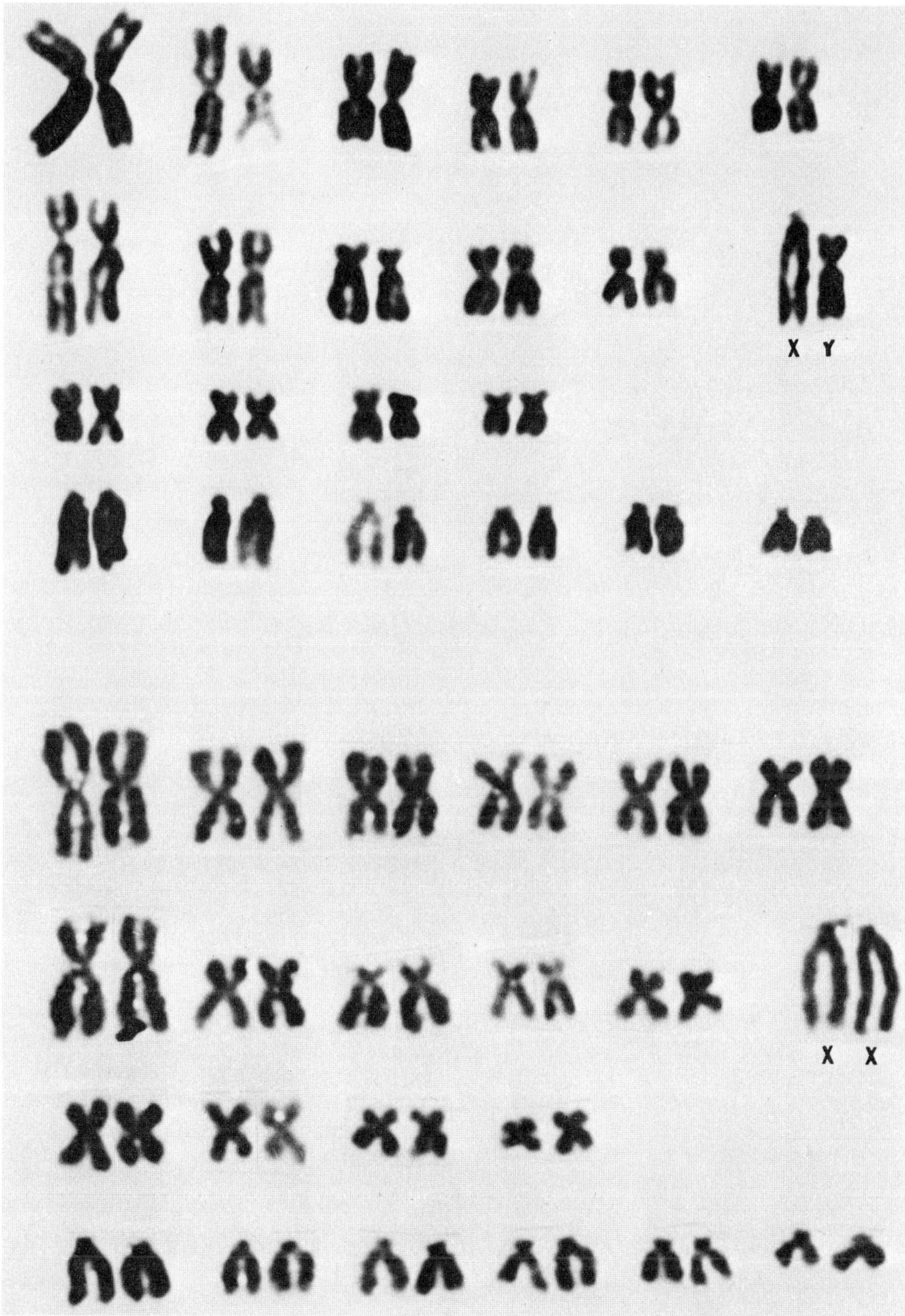

# Order: RODENTIA

## Family: CRICETIDAE

***Meriones shawi*** (Shaw's jird)
2n = 44

Order: RODENTIA Family: CRICETIDAE

*Meriones shawi* (Shaw's jird)

2n=44

AUTOSOMES: 30 Metacentrics and submetacentrics
12 Acrocentrics

SEX CHROMOSOMES: X Submetacentric
Y Submetacentric

The large submetacentric X chromosome is an excellent distinguishing marker when the karyotypes of *M. shawi* are compared with those of *M. libycus* (Folio 226). Several differences also exist among the autosomes between the two species. One of the acrocentric pairs has distinct short arm. This species can be hybridized with *M. libycus*.

The karyotypes presented here are gifts of Charles F. Nadler. The specimens were from laboratory stock originally obtained from Duvernoy-Bahig, Burgel Arab, Egypt.

REFERENCES:


1) Matthey, R.: Les Chromosomes des Muridae. Rev. Suisse de Zool. 60: 225, 1953.

2) Matthey, R.: Cytologie et taxonomie du genre *Meriones*, Illiger. Säugetierk. Mitt. 5:145, 1957.

3) Nadler, C.F. and Lay, D.M.: Chromosomes of some species of *Meriones* (Mammalia: Rodentia). Z. f. Säugetierkunde 32:285, 1967.

4) Lay, D.M. and Nadler, C.F.: Hybridization in the rodent genus *Meriones*. I. Breeding and cytological analyses of *Meriones shawi* (♀) X *Meriones libycus* (♂) hybrids. Cytogenetics 8:35, 1969.

5) Nadler, C.F.: Chromosomal Evolution in Rodents. In "Comparative Mammalian Cytogenetics" (Benirschke, K., ed.), pp. 277-309, Springer-Verlag, New York, 1969.

<u>Meriones shawi</u> (Shaw's jird)

2n=44

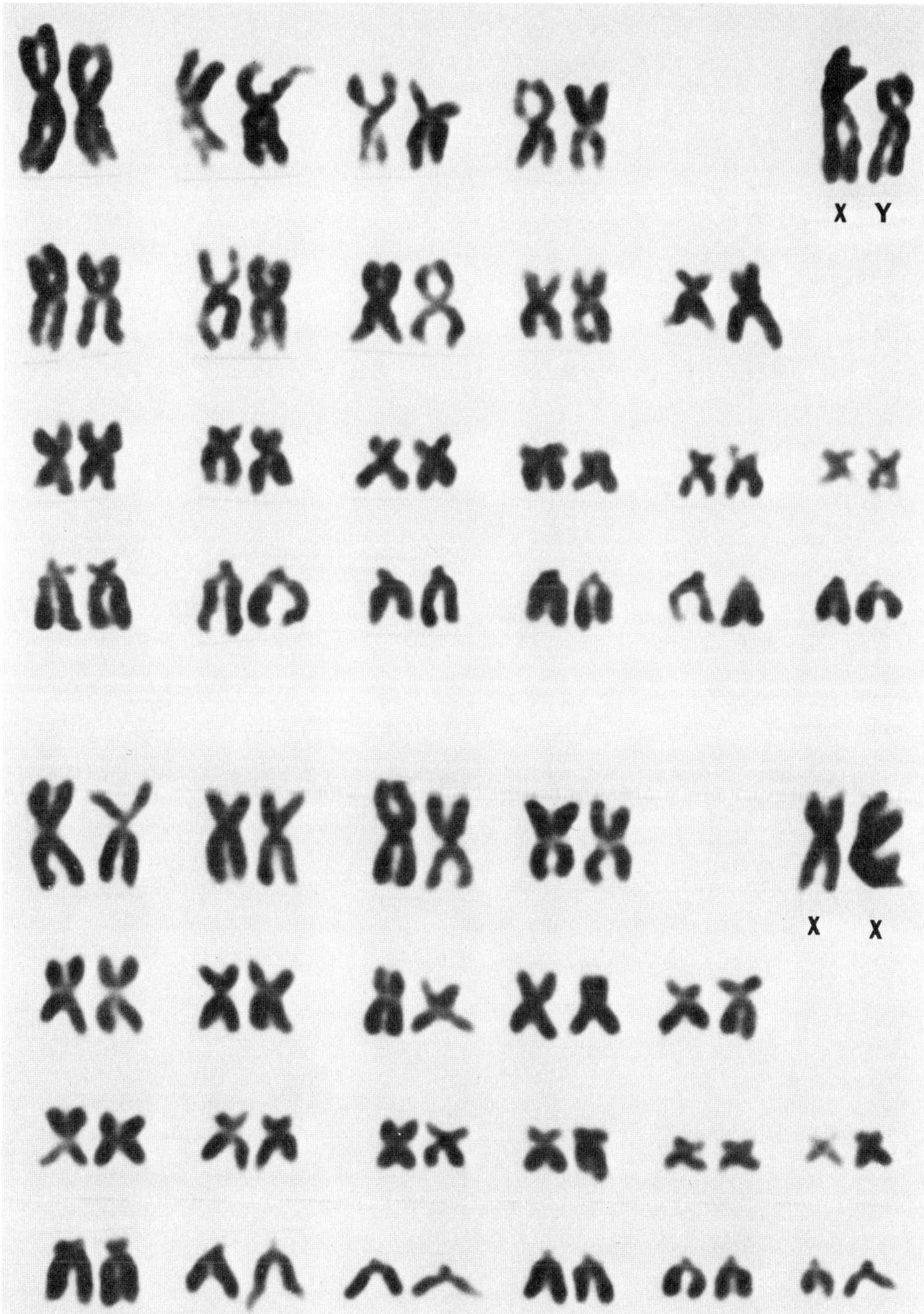

# Order: RODENTIA

## Family: MURIDAE

### *Mus cervicolor*

2n = 40

Order: RODENTIA Family: MURIDAE

## *Mus cervicolor*

2n=40

AUTOSOMES: 38 Acrocentrics

SEX CHROMOSOMES: X Acrocentric
Y Acrocentric

The karyotype of this species is indistinguishable from that of *M. musculus* (Folio 17). Pairing of the chromosomes is arbitrary.

The specimens were laboratory descendents of specimens originally caught in Thailand and were donated by Dr. Frank H. Ruddle, Yale University, New Haven, Connecticut, USA. The subspecies is *M. c. caroli*.

# Mus cervicolor

2n=40

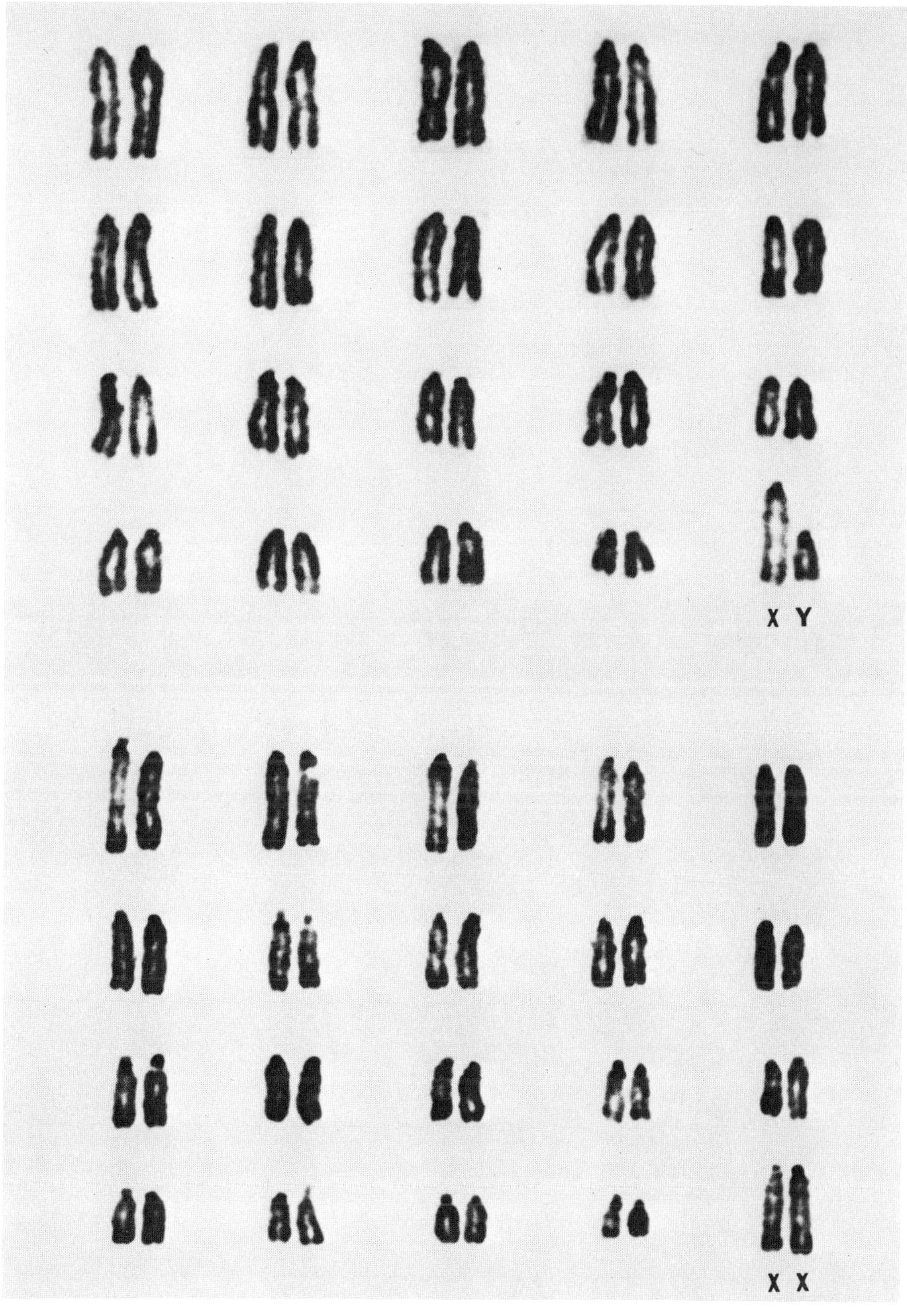

# Order: RODENTIA

## Family: MURIDAE

***Rattus rattus*** (Black rat)

$2n = 42$

Order: RODENTIA Family: MURIDAE

# Rattus rattus (Black rat)

2n=42

AUTOSOMES: 18 Metacentrics and submetacentrics
22 Acrocentrics

SEX CHROMOSOMES: X Acrocentric
Y Acrocentric

All the metacentrics and submetacentrics are either medium-size or small elements. Some of the acrocentrics, especially two pairs, may be long. The X chromosome is among the long acrocentrics, but the identification of the Y offers no problem.

The specimens were collected by Mr. John F. Duncan and his collaborators in South Vietnam. The male was trapped in Calu, and the female, in Quang Tri Combat Base, bearing specimen numbers RHL-58 and DVP 468, respectively. The specimens are stored in the Smithsonian Institute, Washington, D.C., USA. Lung cultures were used for cytological preparations.

Polymorphism exists in many wild populations in the Orient. Yosida and his associates found that the longest autosomal pair and the 9th chromosome may be acrocentric, submetacentric, or hemizygous of these. Also, wild specimens may possess an additional small acrocentric autosome in the complement to give 43 as the diploid number. In the Vietnam samples, such a chromosome was also found in a few individuals.

REFERENCES:

1) Yosida, T.H., Nakamura, A. and Fukaya, T.: Chromosomal polymorphism in Rattus rattus (L.) collected in Kusudomari and Misima. Chromosoma 16:70, 1965.

2) Yosida, T.H. and Associates: Many reports in Ann. Rep. Nat. Inst. Genet., Japan, 1965-1969.

3) Yosida, T.H., Tsuchiya, K., Imai, H.T. and Morwaki, K.: New chromosome types of the black rat, Rattus rattus, collected in Oceania and hybrids between Japanese and Australian rats. Jap. J. Genet. 44:89, 1969.

4) Yosida, T.H., Tsuchiya, K., Imai, H., Moriwaki, K. and Udagawa, T.: Chromosome numbers of rodent species in South East Asia and Oceania. Mammalian Chromosomes Newsletter 10:217, 1969.

5) Yong, H.S.: Karyotypes of rats from Hong Kong and Thailand (Muridae, genus Rattus Fischer). Cytologia 34:394, 1969.

6) Yong, H.S.: Karyotypes of Malayan rats (Rodentia-Muridae, genus Rattus Fischer). Chromosoma 27:245, 1969.

7) Capanna, E. and Ciritelli, M.V.: An endemic population of Rattus rattus (L.) with a 38-chromosome complement. Mammalian Chromosomes Newsletter 10: 220, 1969.

Rattus rattus (Black rat)

2n=42

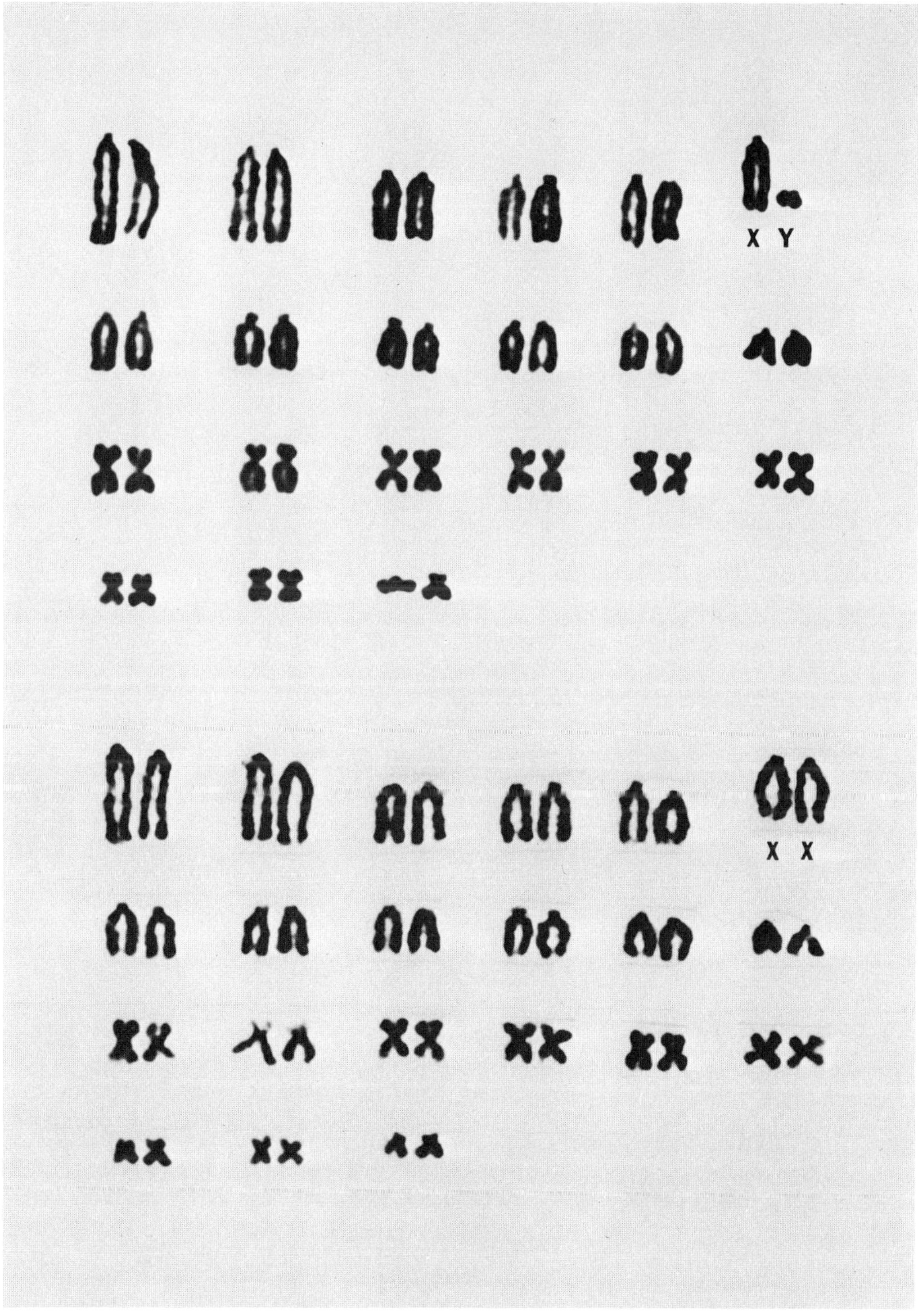

# Order: CARNIVORA

## Family: CANIDAE

*Nyctereutes procyonoides (viverrinus)* (Raccoon dog)
$2n = 42$

Order: CARNIVORA Family: CANIDAE

<u>Nyctereutes</u> <u>procyonoides</u> (<u>viverrinus</u>) (Raccoon dog)

2n=42

AUTOSOMES: 26 Metacentrics and submetacentrics
14 Acrocentrics

SEX CHROMOSOMES: X Acrocentric
Y Acrocentric with satellites

Skin biopsies of a male and a female were kindly provided by Dr. C. Gray, National Zoological Park, Washington, D.C., USA. The male was mosaic for 42/43 chromosomes, apparently XY/XXY, although testicular biopsy disclosed normal seminiferous tubules. The X chromosome was identified by autoradiography.

REFERENCES:

1) Minouchi, O.: On the spermatogenesis of the raccoon dog (<u>Nyctereutes viverrinus</u>), with special reference to the sex chromosomes. Cytologia <u>1</u>: 88, 1929.

2) Wurster, D.H.: Cytogenetic and phylogenetic studies in Carnivora. In <u>Comparative Mammalian Cytogenetics</u> (Benirschke, K., ed.), Springer-Verlag, New York, 1969.

3) Todd, N.B. and Pressman, S.R.: The karyotype of the raccoon dog (<u>Nyctereutes</u> sp.). Mammalian Chromosomes Newsletter <u>10</u>:21, 1969.

Nyctereutes procyonoides (viverrinus) (Raccoon dog)

2n=42

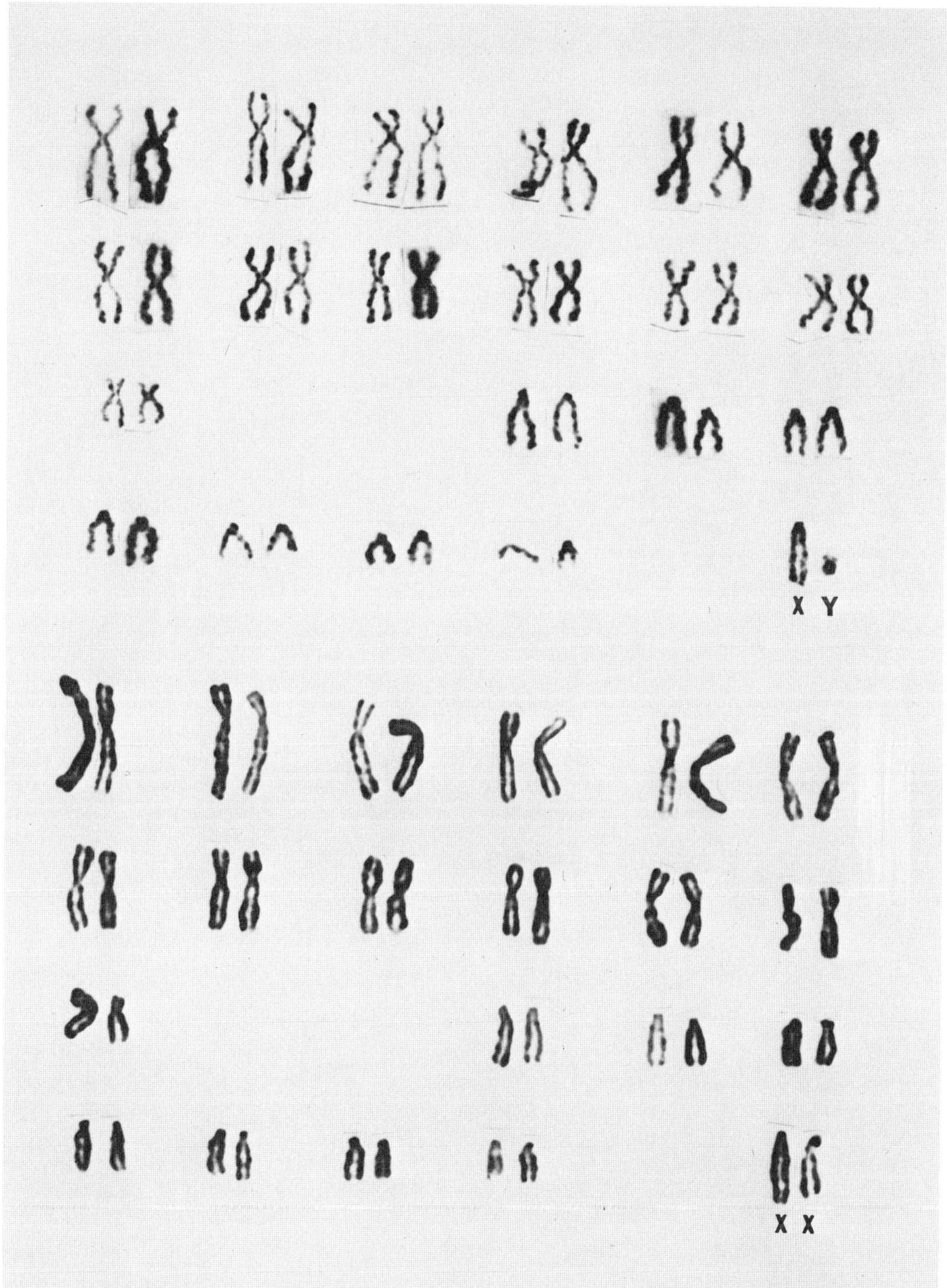

# Order: CARNIVORA

## Family: MUSTELIDAE

***Mustela frenata*** (Long-tailed weasel)
2n = 42

Order: CARNIVORA                                    Family: MUSTELIDAE

*Mustela frenata* (Long-tailed weasel)

2n=42

AUTOSOMES: 20 Metacentrics and submetacentrics
20 Acrocentrics

SEX CHROMOSOMES: X Submetacentric
Y Submetacentric

Metaphases for these karyotypes were kindly donated by Dr. P. K. Basrur, Guelph, Ontario, Canada, and came from blood and fibroblast cultures of three male and two female animals trapped in Southern Ontario. The animals were identified by features such as skull measurements, and Basrur gives the measurements of arm ratios. She considers there to be only 18 acrocentrics, the second pair to last autosomes shown here were considered as biarmed elements. A secondary constriction is regularly found near the ends of the long arms of the last element in the second row and less commonly in the same position of the elements shown here in position two.

REFERENCES:

1) Basrur, P.K.: The karyotype of the long-tailed weasel *Mustela frenata noveboracensis* Emmons. Canad. J. Genet. Cytol. 10:390, 1968.

<u>Mustela</u> <u>frenata</u> (Long-tailed weasel)

2n=42

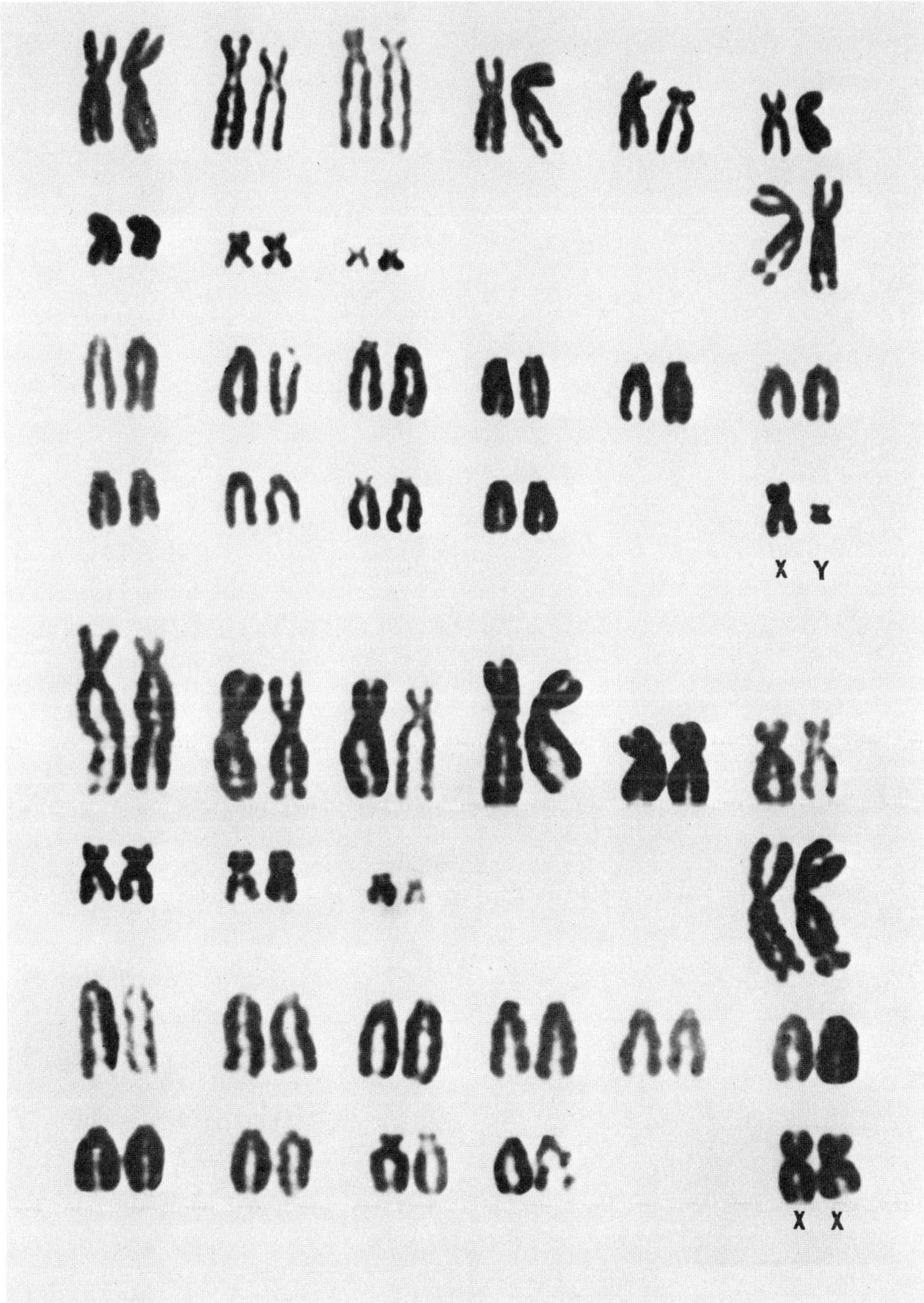

# Order: CARNIVORA

## Family: VIVERRIDAE

***Bdeogale sp.*** (Black-footed mongoose)

2n = 36

Order: CARNIVORA Family: VIVERRIDAE

*Bdeogale sp.* (Black-footed mongoose)

2n=36

AUTOSOMES: 28 Metacentrics and submetacentrics
6 Acrocentrics

SEX CHROMOSOMES: X Submetacentric
Y Submetacentric

The size of the short arms of No. 5 appears to be variable. No chromosome pair showed conspicuous secondary constrictions found as good marker chromosomes in many carnivore karyotypes.

Skin biopsies were kindly made available from one animal of each sex by Dr. C. Gray, National Zoological Park, Washington, D.C., USA.

REFERENCES:

1) Wurster, D.H. and Benirschke, K.: Chromosome numbers in thirty species of carnivores. Mammalian Chromosomes Newsletter 8:195, 1967.

2) Wurster, D.H. and Benirschke, K.: Comparative cytogenetic studies in the Order Carnivora. Chromosoma 24:336, 1968.

<u>Bdeogale</u> <u>sp</u>. (Black-footed mongoose)

2n=36

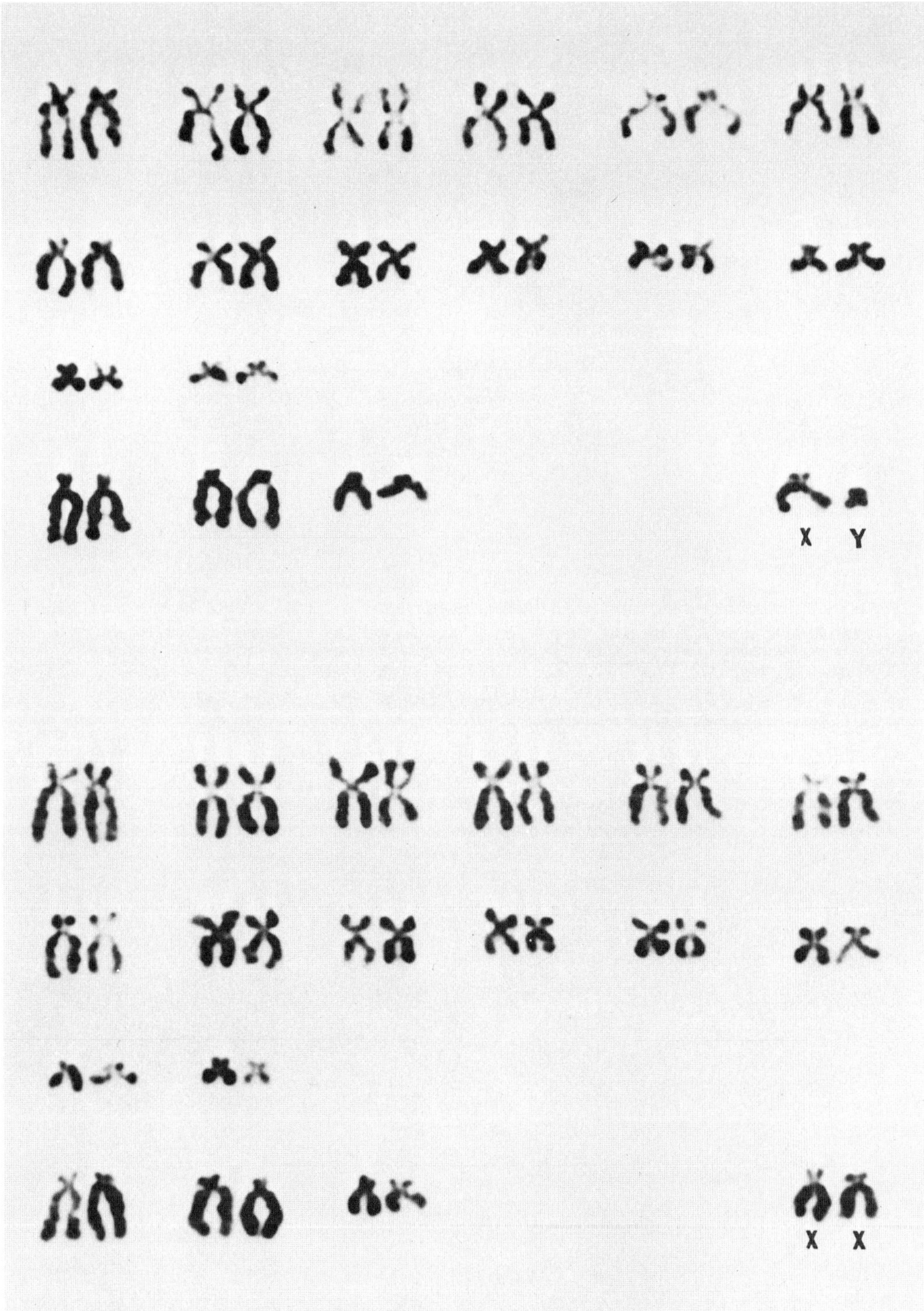

# Order: CARNIVORA

## Family: VIVERRIDAE

***Paradoxurus hermaphroditus cochinensis*** (Palm civet)

2n = 42

Order: CARNIVORA Family: VIVERRIDAE

*Paradoxurus hermaphroditus cochinensis* (Palm civet)

2n=42

AUTOSOMES: 22 Metacentrics and submetacentrics
18 Acrocentrics

SEX CHROMOSOMES: X Submetacentric
Y Minute acrocentric

The marker chromosome pair with pronounced secondary constrictions and satellites is here placed as element No. 7.

Skin biopsies of one sex each were kindly provided by Mr. J. F. Duncan from animals captured in South Vietnam.

REFERENCES:

1) Ray-Chaudhuri, S.P., Ranjini, P.V. and Sharma, T.: Somatic chromosomes of the common palm civet, *Paradoxurus hermaphroditus*. Experientia 22:740, 1966.

<u>Paradoxurus hermaphroditus cochinensis</u> (Palm civet)

2n=42

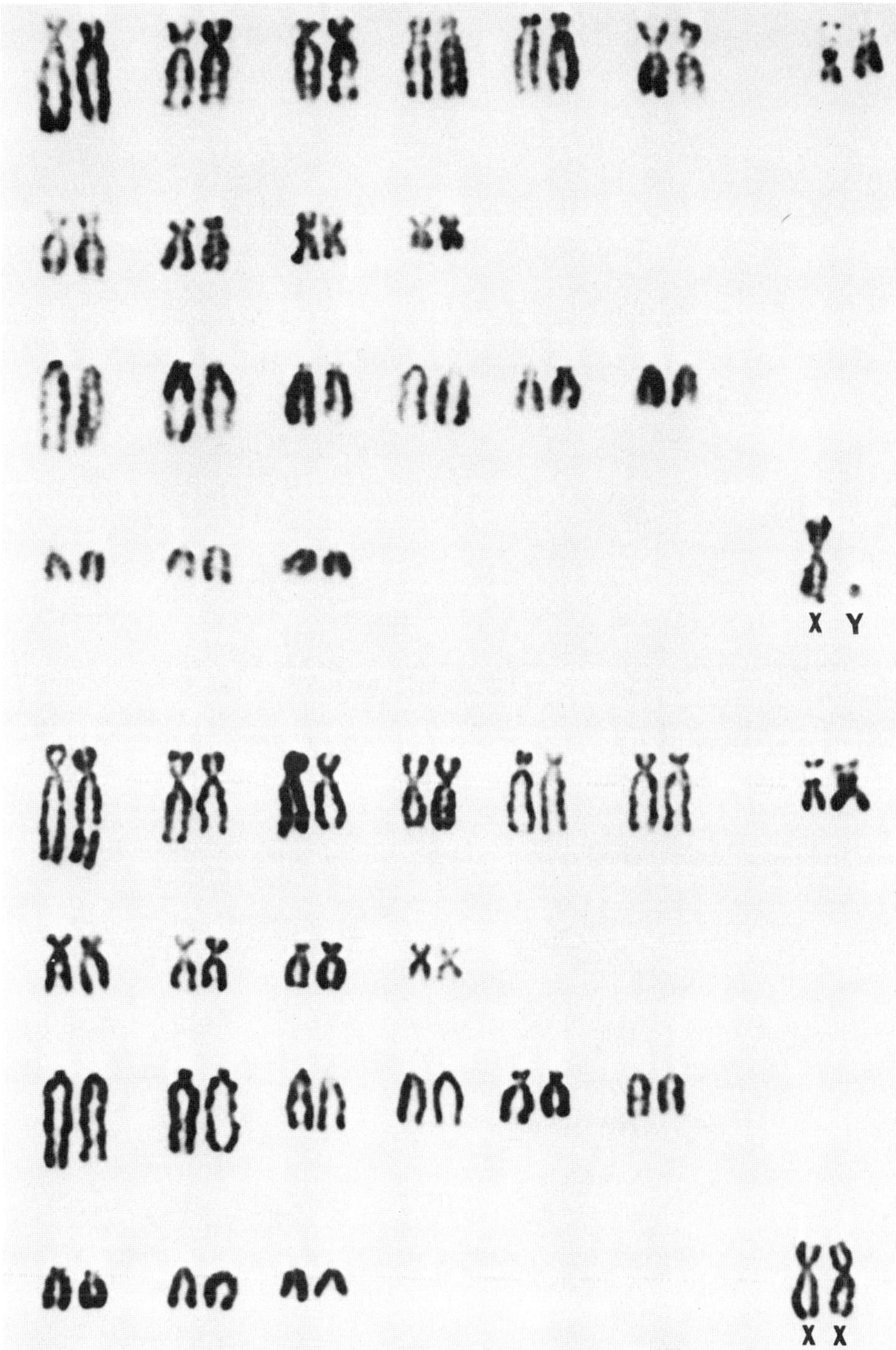

# Order: CARNIVORA

## Family: FELIDAE

***Acinonyx jubatus*** (Cheetah)

2n = 38

Volume 5, Folio 234, 1971

Order: CARNIVORA Family: FELIDAE

*Acinonyx jubatus* (Cheetah)

2n=38

AUTOSOMES: 34 Metacentrics and submetacentrics
2 Acrocentrics

SEX CHROMOSOMES: X Submetacentric
Y Acrocentric

According to the San Juan nomenclature system for Felidae (see Folio 31) the autosomes are composed of three pairs in group A, four pairs in group B, two pairs in group C, four pairs in group D, four pairs in group E, and one pair in group F. The satellited pair is $E_1$.

The male karyotype was donated by Dr. W. L. Hard, Omaha, Nebraska, USA, and was prepared from a lymphocyte culture of an animal imported from Somaliland and located at H. Doorley Zoo, Omaha. The female karyotype came from skin fibroblast cultures taken by Dr. C. Gray, National Zoological Park, Washington, D.C., USA.

REFERENCES:

1) Hsu, T.C., Rearden, H.H. and Luquette, G.F.: Karyological studies of nine species of Felidae. Amer. Naturalist 97:225, 1963.

2) Hard, W.L.: The karyotype of a male cheetah, *Acinonyx jabatus jabatus*. Mammalian Chromosomes Newsletter 9:16, 1968.

Acinonyx jubatus (Cheetah)

2n=38

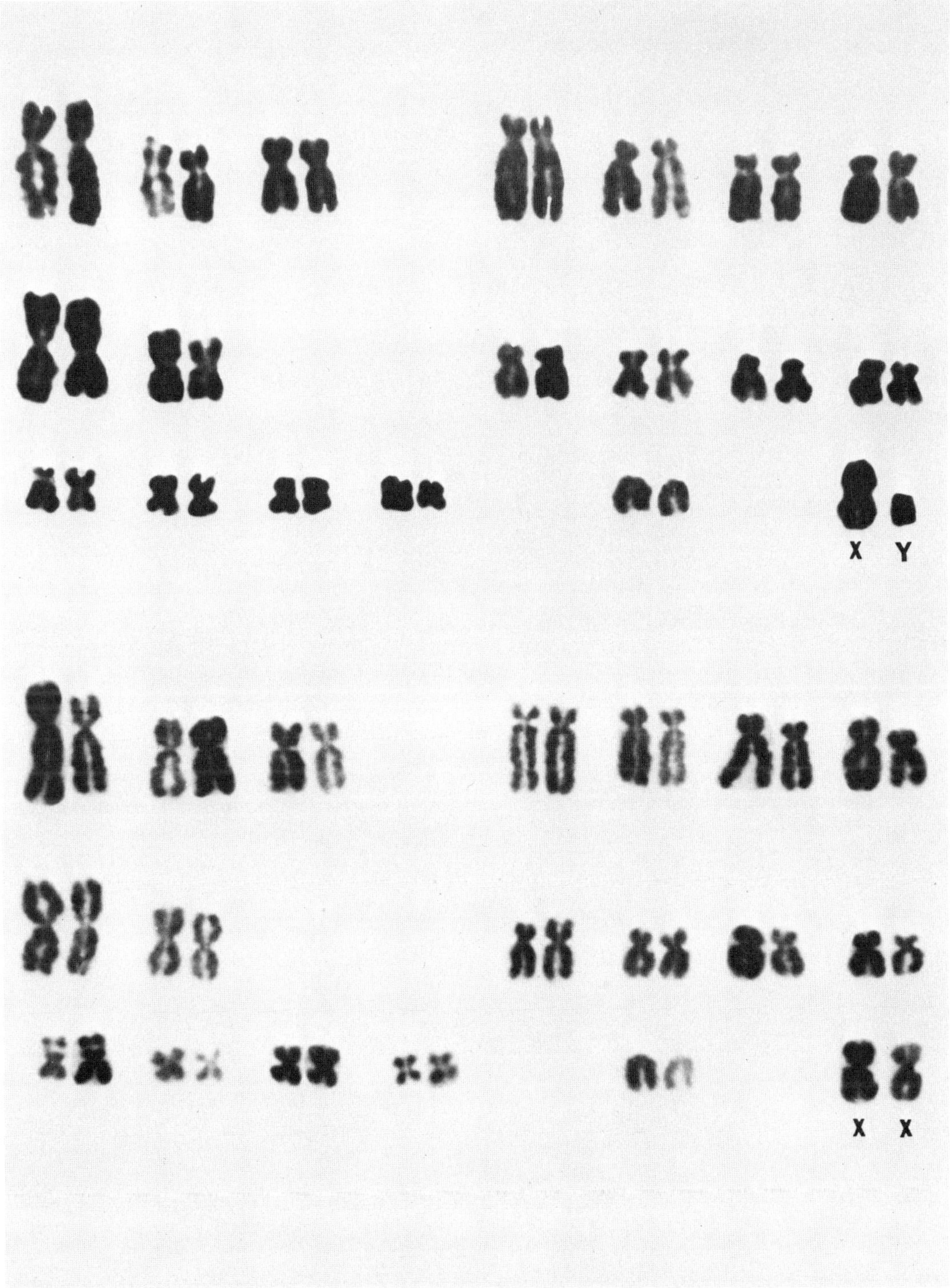

# Order: CARNIVORA

## Family: FELIDAE

***Felis serval*** (Serval)

2n = 38

Order: CARNIVORA Family: FELIDAE

*Felis serval* (Serval)

2n=38

AUTOSOMES: 32 Metacentrics and submetacentrics
4 Acrocentrics

SEX CHROMOSOMES: X Metacentric
Y Submetacentric

Skin biopsies of these two animals were kindly provided by Dr. C. Gray, National Zoological Park, Washington, D.C., USA. The karyotype is arranged following the San Juan Convention for Felidae (see Folio 31). The marker chromosome is the first in row five ($E_1$).

REFERENCES:


1) Wurster, D.H. and Benirschke, K.: Karyotypes of four more species of cats. Mammalian Chromosomes Newsletter 9:236, 1968.

2) Wurster, D.H. and Benirschke, K.: Comparative cytogenetic studies in the Order Carnivora. Chromosoma 24:336, 1968.

# Felis serval (Serval)

2n=38

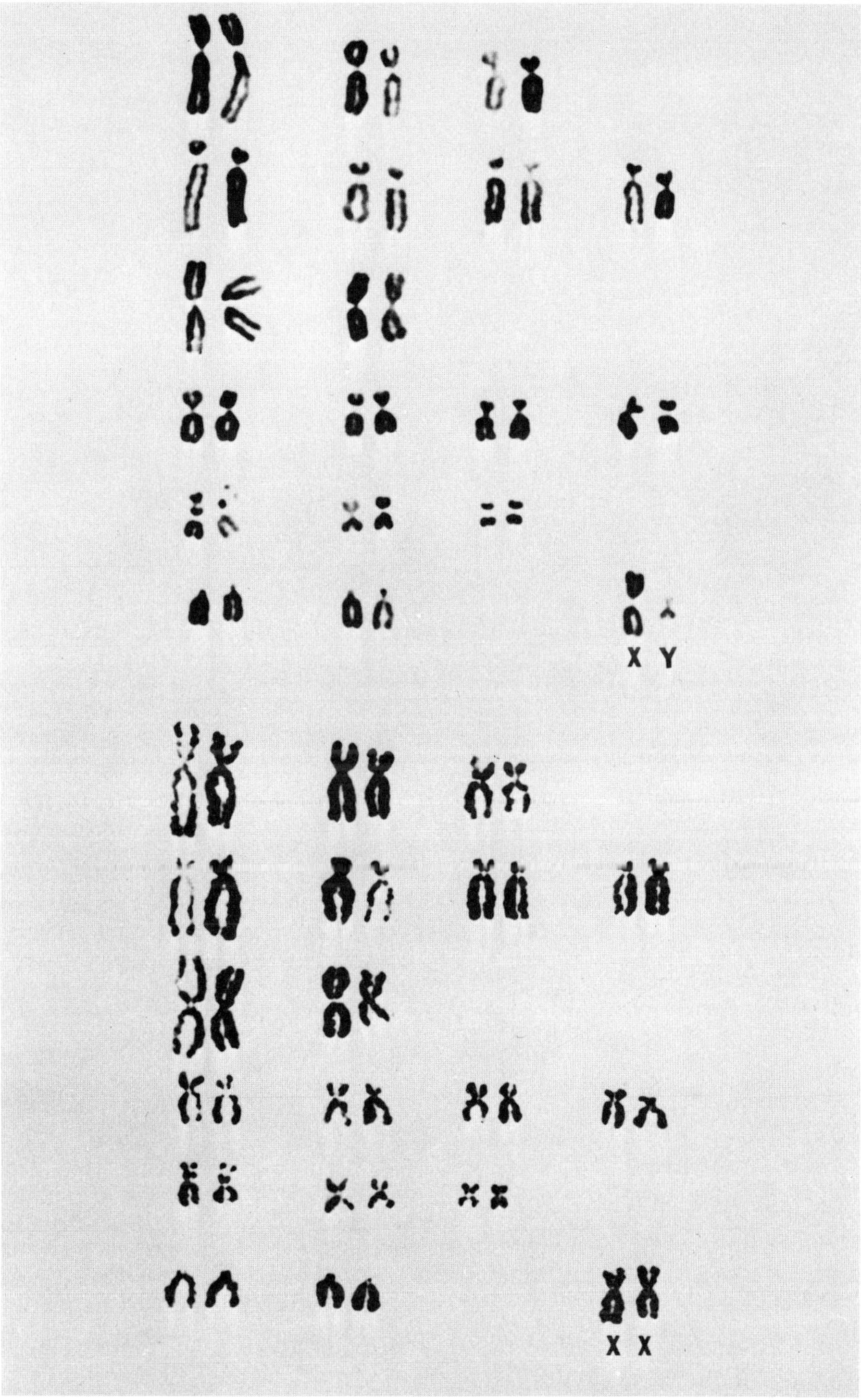

# Order: CARNIVORA

## Family: FELIDAE

***Felis viverrina*** (Fishing cat)
$2n = 38$

Order: CARNIVORA Family: FELIDAE

# *Felis viverrina* (Fishing cat)

2n=38

AUTOSOMES: 34 Metacentrics and submetacentrics
2 Acrocentrics

SEX CHROMOSOMES: X Submetacentric
Y Submetacentric

Skin biopsies of both animals were kindly provided by Dr. C. Gray, National Zoological Park, Washington, D.C., USA. The karyotype is arranged according to the San Juan Convention for Felidae (see Folio 31). The marker chromosome ($E_1$) is the first element in the last row of each karyotype.

REFERENCES:

1) Wurster, D.H. and Benirschke, K.: The chromosomes of three species of cats (*Felis nigripes*, *F. bengalensis*, *F. viverrina*). Mammalian Chromosomes Newsletter 9:20, 1968.

2) Wurster, D.H. and Benirschke, K.: Comparative cytogenetic studies in the Order Carnivora. Chromosoma 24:336, 1968.

3) Leyhausen, P. and Tonkin, B.A.: Comment on the karyotypes of the leopard cat and the fishing cat. Mammalian Chromosomes Newsletter 9:78, 1968.

<u>Felis</u> <u>viverrina</u> (Fishing cat)

2n=38

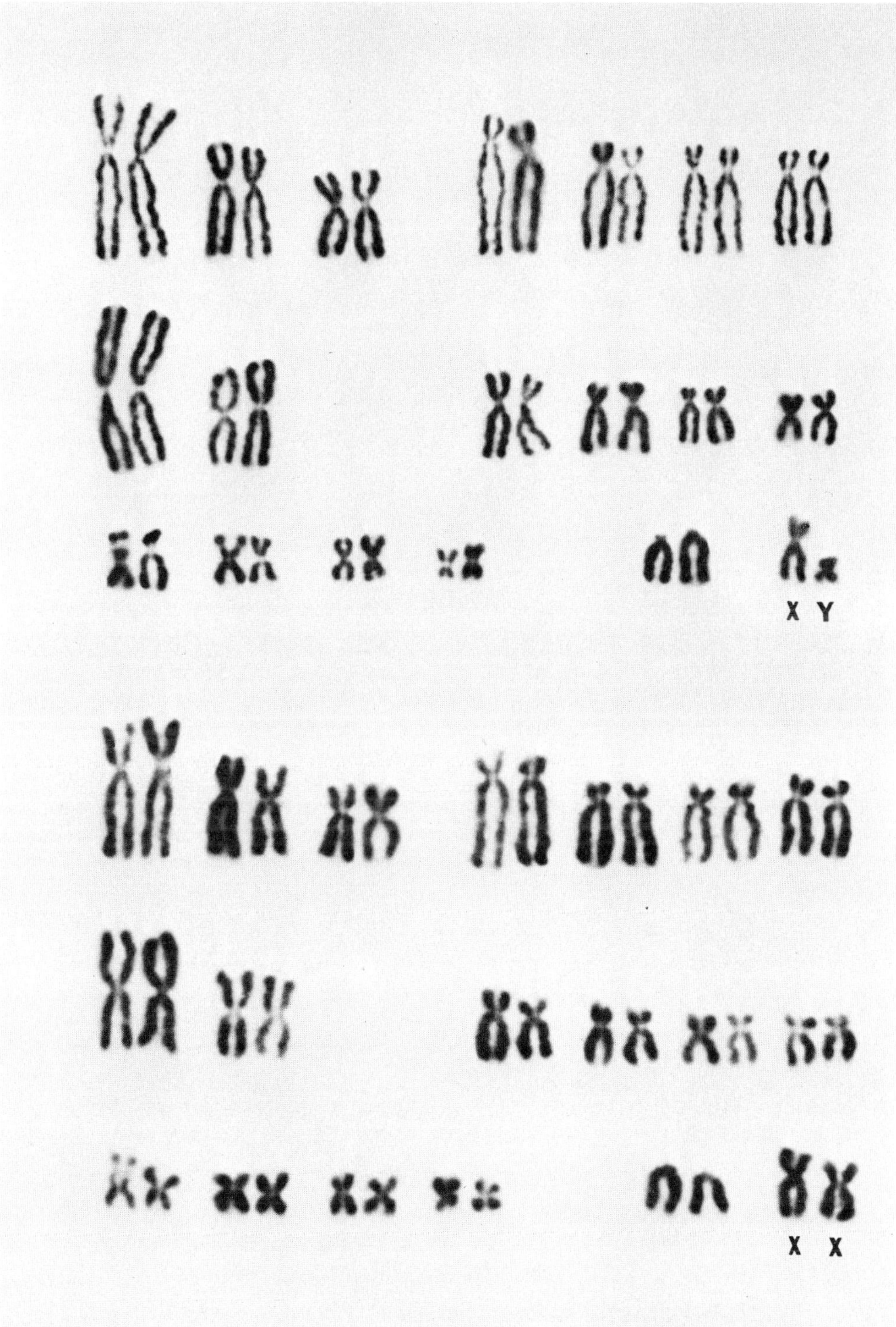

# Order: CARNIVORA

## Family: FELIDAE

***Felis (Leopardus) wiedi(i)*** (Marguay cat)
2n = 36

Order: CARNIVORA Family: FELIDAE

Felis (Leopardus) wiedi(i) (Marguay cat)

2n=36

AUTOSOMES: 32 Metacentrics and submetacentrics
2 Acrocentrics

SEX CHROMOSOMES: X Submetacentric
Y Submetacentric

A skin biopsy of a male specimen was kindly made available by Dr. C. Gray of the National Zoological Park, Washington, D.C., USA. The female karyotype was donated by Dr. R. Richart, New York, USA, and came from an animal at Bronx Zoo. The karyotypes are arranged according to the San Juan Convention for Felidae (see Folio 31). The marker is placed as the first element in row five ($E_1$).

In the report of Hsu, Rearden and Luquette, the karyotype of F. wiedii contained no acrocentric pairs. Leyhausen pointed out that the karyotype matches that of F. tigrinus (little spotted cat). It is possible that the specimen used in Hsu's study was misidentified and was actually F. tigrinus. True F. wiedi has one pair of acrocentric autosomes.

REFERENCES:

1) Hsu, T.C.: Two species of cats with 36 chromosomes. Mammalian Chromosomes Newsletter No. 8:4, 1962.

2) Hsu, T.C., Rearden, H. and Luquette, G.F.: Karyological studies of nine species of Felidae. Amer. Naturalist 97:225, 1963.

3) Leyhausen, P.: The karyotypes of two cat species. Mammalian Chromosomes Newsletter 8:287, 1967.

Felis (Leopardus) wiedi(i) (Marguay cat)

2n=36

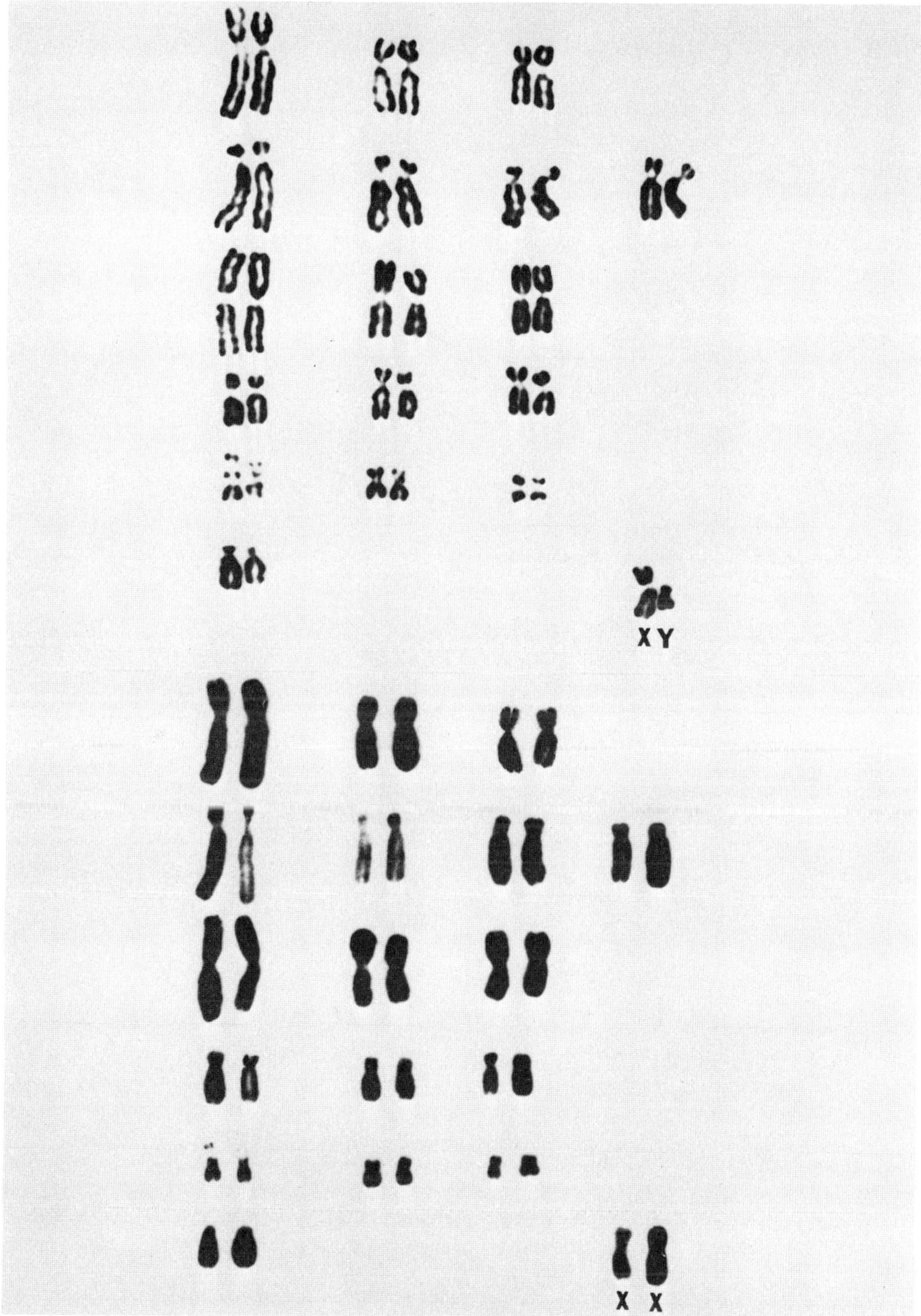

# Order: TUBULIDENTATA

## Family: ORYCTEROPODIDAE

***Orycteropus afer*** (Aardvark)

2n = 20

Order: TUBULIDENTATA    Family: ORYCTEROPODIDAE

## *Orycteropus afer* (Aardvark)

2n=20

AUTOSOMES: 18 Metacentrics and submetacentrics

SEX CHROMOSOMES: X Metacentric
Y Metacentric

The two largest autosomes often have narrow secondary constrictions which correspond to areas of late replication. The Y chromosome labels very late and heavily. The X chromosome constitutes approximately 5% of the haploid set. Dr. N. B. Atkin, Northwood, England, has measured the DNA content of cultured cells. The DNA content of these cells is approximately 1.6 x that of human lymphocytes.

The male skin biopsy came from an animal at the National Zoological Park, Washington, D.C., USA, through the courtesy of Dr. C. Gray. The female karyotype came from a skin biopsy of an animal at Tacoma Zoo, Tacoma, Washington, USA, supplied by Dr. M. L. Johnson. A second female specimen from Crandon Park Zoo, Miami, Florida, USA, gave identical results.

REFERENCES:

1) Wurster, D. and Benirschke, K.: The chromosomes of the aardvark, *Orycteropus afer*. Mammalian Chromosomes Newsletter 11:34, 1970.

Order: TUBULIDENTATA Family: ORYCTEROPODIDAE

Orycteropus afer (Aardvark)

2n=20

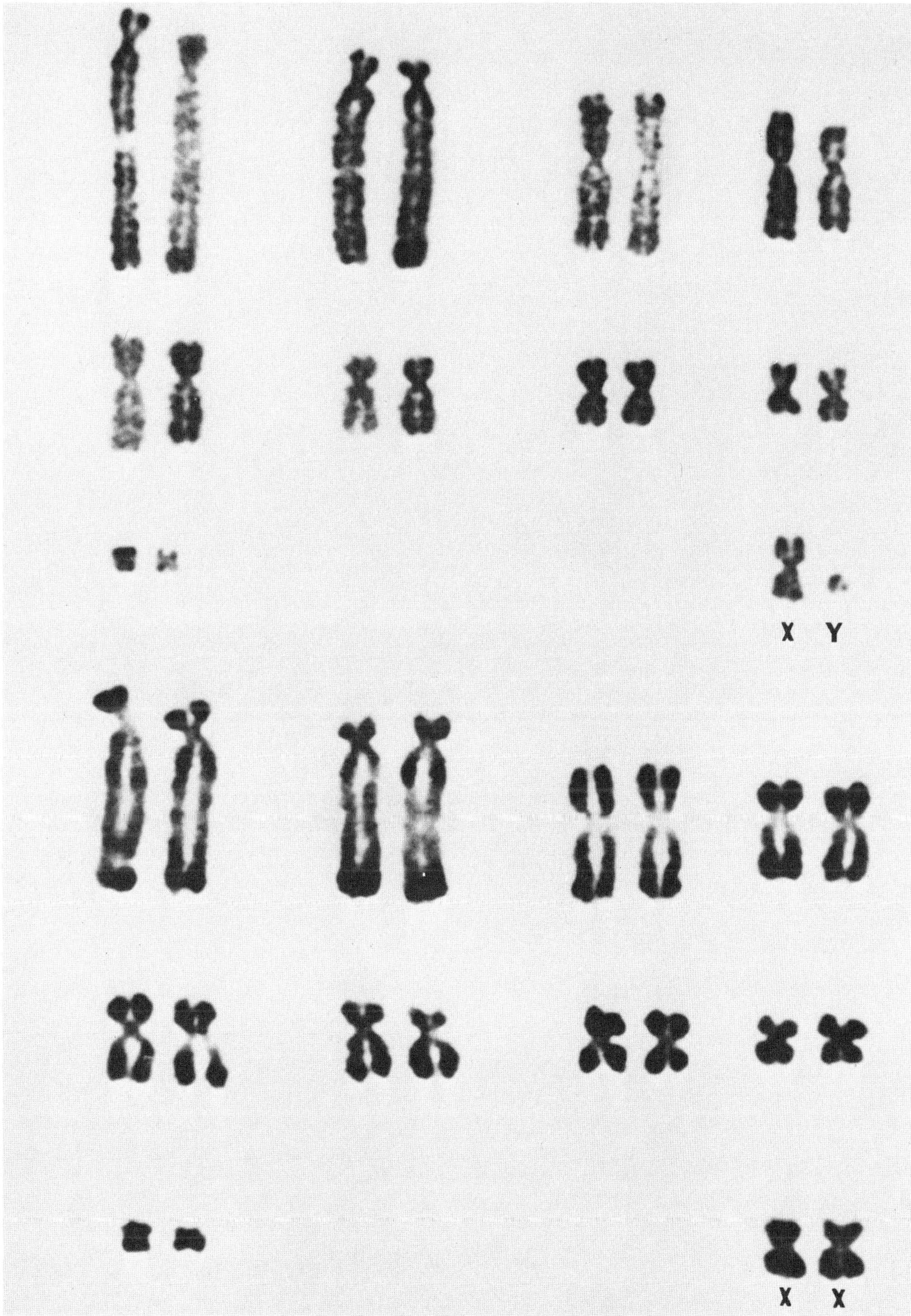

# Order: PROBOSCIDEA

## Family: ELEPHANTIDAE

*Elephas maximus* (Indian or Asian elephant)
$2n = 56$

Order: PROBOSCIDEA Family: ELEPHANTIDAE

*Elephas maximus* (Indian or Asian elephant)

2n=56

AUTOSOMES: 12 Metacentrics and submetacentrics
42 Acrocentrics

SEX CHROMOSOMES: X Submetacentric
Y Submetacentric

Among the biarmed autosomes, two pairs (the 2nd and the 3rd in row 1) may be classified as acrocentrics by some cytologists. In that case, the metacentric and submetacentric autosomes are reduced to 8. The Y chromosome may be called a subtelocentric by some authors.

The karyotypes presented here are gifts of Drs. David A. Hungerford (Institute for Cancer Research, Philadelphia, Pennsylvania, USA) and R. L. Snyder (Penrose Research Laboratory, Zoological Society of Philadelphia, Pennsylvania, USA). The specimens were display animals of the Philadelphia Zoo. Another set of karyotypes donated to us by Dr. H. S. Norberg, Norway showed similar results. All were from cell cultures. We would like to thank Dr. Norberg for his kindness.

REFERENCES:

1) Hungerford, D. A., Chandra, H. S., Snyder, R. L. and Ulmer, F. A.: Chromosomes of three elephants, two Asian (*Elephas maximus*) and one African (*Loxodonta africana*). *Cytogenetics* 5:243, 1966.

2) Norberg, H. S.: The chromosomes of the Indian female elephant (*Elephas indicus* syn. *E. maximus* L.). Hereditas *63*:279, 1969.

Elephas maximus (Indian or Asian elephant)

2n=56

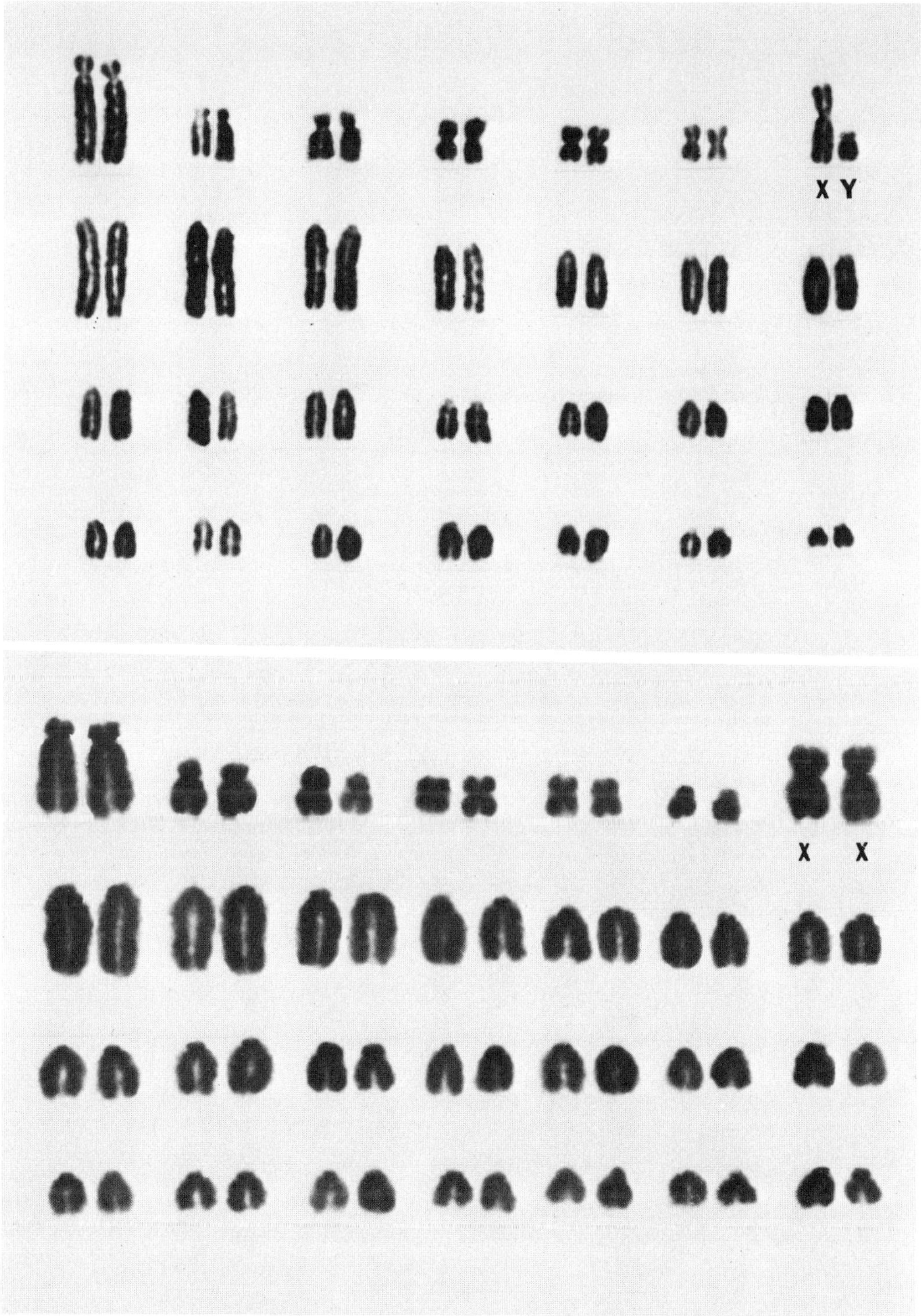

# Order: PERISSODACTYLA

## Family: EQUIDAE

### *Equus quagga* (Burchell's zebra, Plains zebra)

2n = 44

Order: PERISSODACTYLA  Family: EQUIDAE

*Equus quagga* (Burchell's zebra, Plains zebra)

2n=44

AUTOSOMES: 34 Metacentrics and submetacentrics
8 Acrocentrics

SEX CHROMOSOMES: X Metacentric
Y Minute ?acrocentric

Some of the smaller autosomes are difficult to classify, hence the species is often considered to possess only 6 acrocentric autosomes. Also, the Y is so small that it is not certain to be acrocentric.

Several different subspecies of this common zebra have been studied, all with similar results. The male karyotype shown here is a Damara zebra (*E. q. antiquorum*), the female is of the rare "maneless" mutant (*E. q. boehmi*). Skin biopsies were kindly provided by Dr. H. Heck, Catskill Game Farm, New York, USA, who also identified the species.

REFERENCES:

1) Benirschke, K. and McFeeley, R.A.: The chromosomes of the Grant zebra, *Equus quagga boehmi*. Mammalian Chromosomes Newsletter No. 10:82, 1963.

2) Mutton, D.E., King, J.M. and Hamerton, J.L.: Chromosome studies in the genus *Equus*. Mammalian Chromosomes Newsletter No. 13:7, 1964.

3) Benirschke, K. and Malouf, N.: Chromosome studies of Equidae. In *Equus*, Vols. 1 & 2 (H. Dathe, ed.), Tierpark Berlin, 253, 1967.

4) King, J.M., Short, R.V., Mutton, D.E. and Hamerton, J.L.: The reproductive physiology of male zebra-horse and zebra-donkey hybrids. J. Reprod. Fertil. 9:391, 1965.

*Equus quagga* (Burchell's zebra, Plains zebra)

2n=44

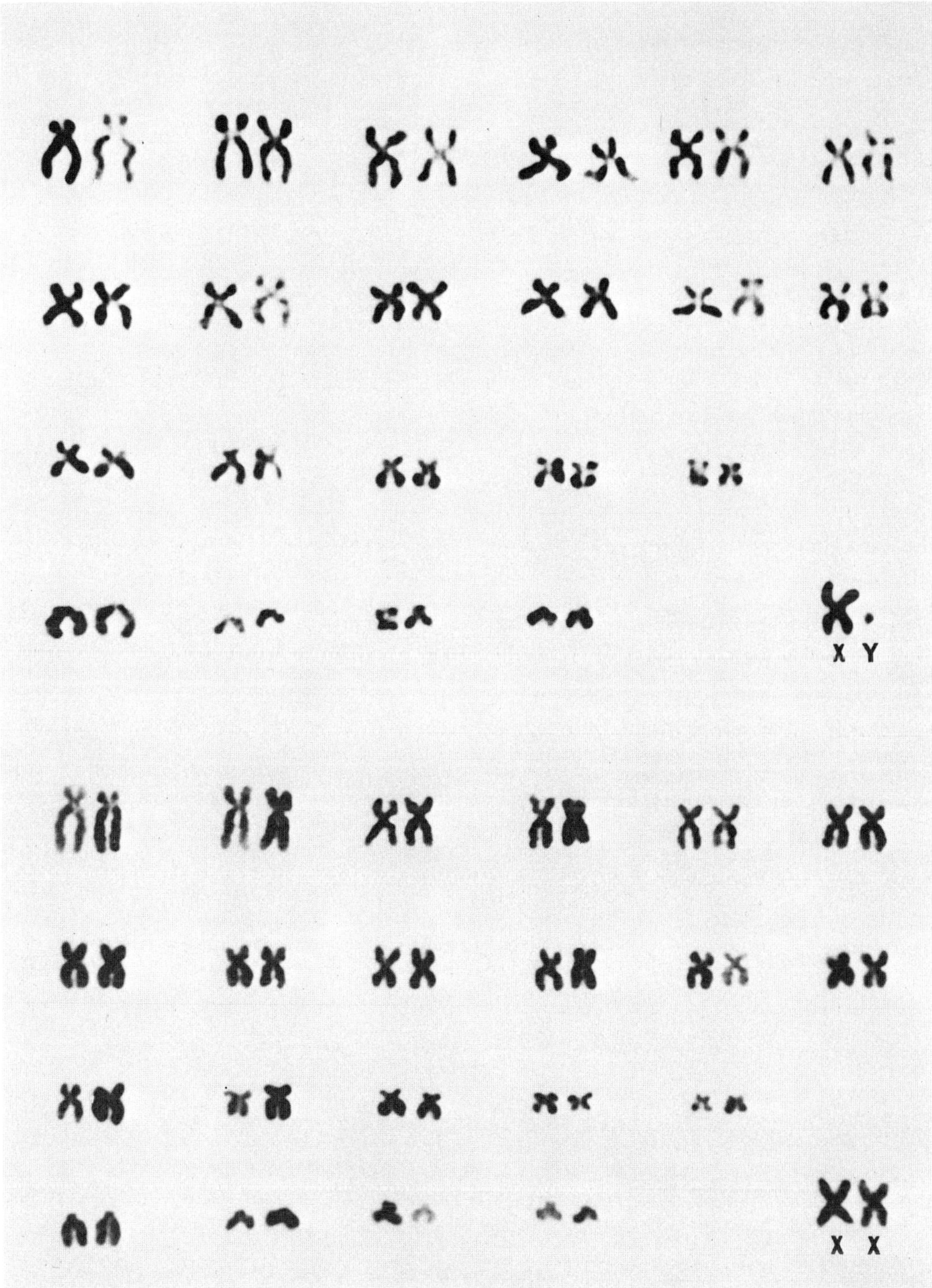

# Order: ARTIODACTYLA

## Family: CERVIDAE

***Elaphurus davidianus*** (Pere David's deer)

$2n = 68$

Order: ARTIODACTYLA Family: CERVIDAE

*Elaphurus davidianus* (Père David's deer)

2n=68

AUTOSOMES: 2 Metacentrics
64 Acrocentrics

SEX CHROMOSOMES: X Acrocentric
Y Metacentric

Skin biopsies were kindly made available of a male by Dr. C. Gray, National Zoological Park, Washington, D.C., USA, and of a female by Dr. H. Heck, Catskill Game Farm, New York, USA.

*Elaphurus davidianus* (Père David's deer)

2n=68

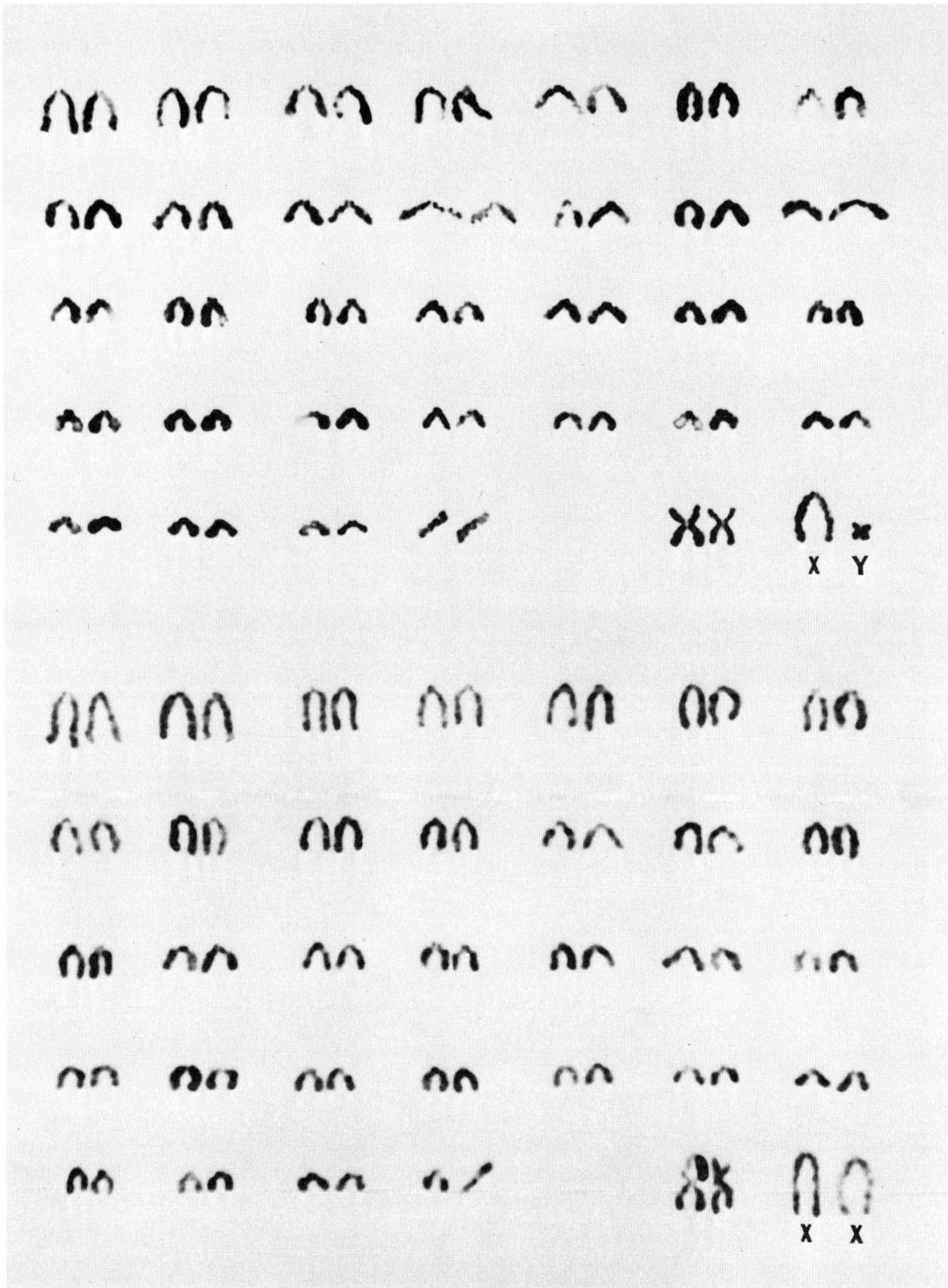

# Order: ARTIODACTYLA

## Family: BOVIDAE

*Cephalophus silvicultor* (Yellow-backed duiker)
2n = 60

Order: ARTIODACTYLA Family: BOVIDAE

Cephalophus silvicultor (Yellow-backed duiker)

2n=60

AUTOSOMES: 58 Acrocentrics

SEX CHROMOSOMES: X Submetacentric
Y Acrocentric

Blood cultures of one male and one female animal were used for the preparation of these karyotypes. They were kindly donated by Dr. W. L. Hard, Omaha, Nebraska, USA, and came from animals at the Henry Doorley Zoo, Omaha.

While the identification of the X is unequivocal, the Y is not positively identified. Several other duikers have similar karyotypes as discussed by Hard.

REFERENCES:

1) Hard, W.L.: The chromosomes of duikers. Mammalian Chromosomes Newsletter 10:216, 1969.

<u>Cephalophus</u> <u>silvicultor</u> (Yellow-backed duiker)

2n=60

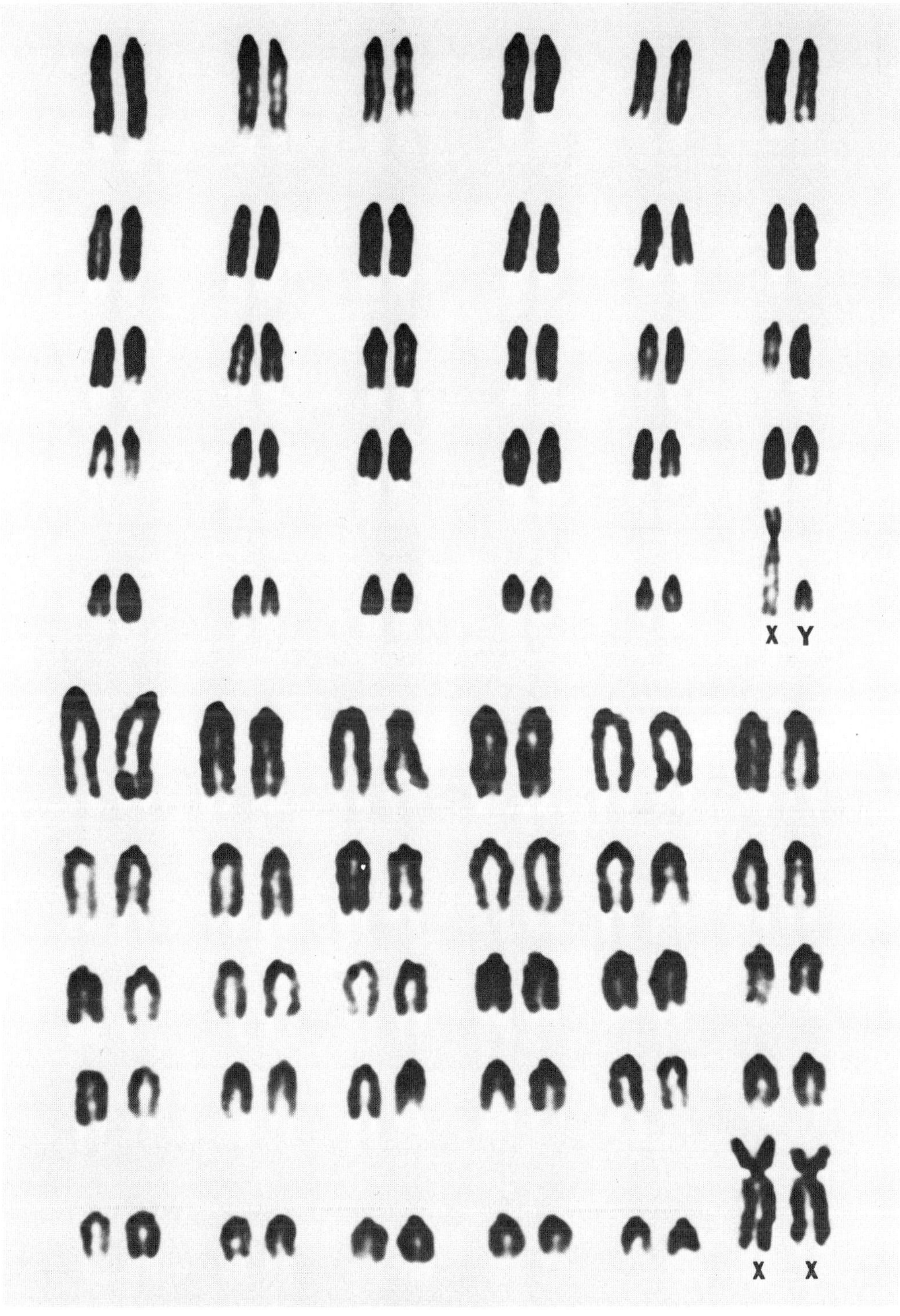

# Order: ARTIODACTYLA

## Family: BOVIDAE

### *Rupicapra rupicapra* (Chamois)
2n = 58

Order: ARTIODACTYLA　　　　　　　　　　　　　　　　　　　　　　　　Family: BOVIDAE

*Rupicapra rupicapra* (Chamois)

2n=58

AUTOSOMES: 2 Metacentrics
54 Acrocentrics

SEX CHROMOSOMES: X Acrocentric
Y Metacentric

These karyotypes, prepared from kidney cultures, were kindly donated by Dr. A. Gropp, Bonn, Germany. Three animals were studied from Valais, Switzerland. The X was identified by autoradiography. The paracentromeric region of the large metacentric is late replicating.

REFERENCES:


1) Gropp, A., Giers, D. and Wandeler, A.: The karyotype of the chamois (*Rupicapra rupicapra*, Bovidae). Mammalian Chromosomes Newsletter 10:19, 1969.

2) Gropp, A., Giers, D., Fernandez-Donoso, R., Tiepolo, L. and Fraccaro, M.: The chromosomes of the chamois (tribe *Rupicaprini* Simpson). Cytogenetics 9:1, 1970.

Rupicapra rupicapra (Chamois)

2n=58

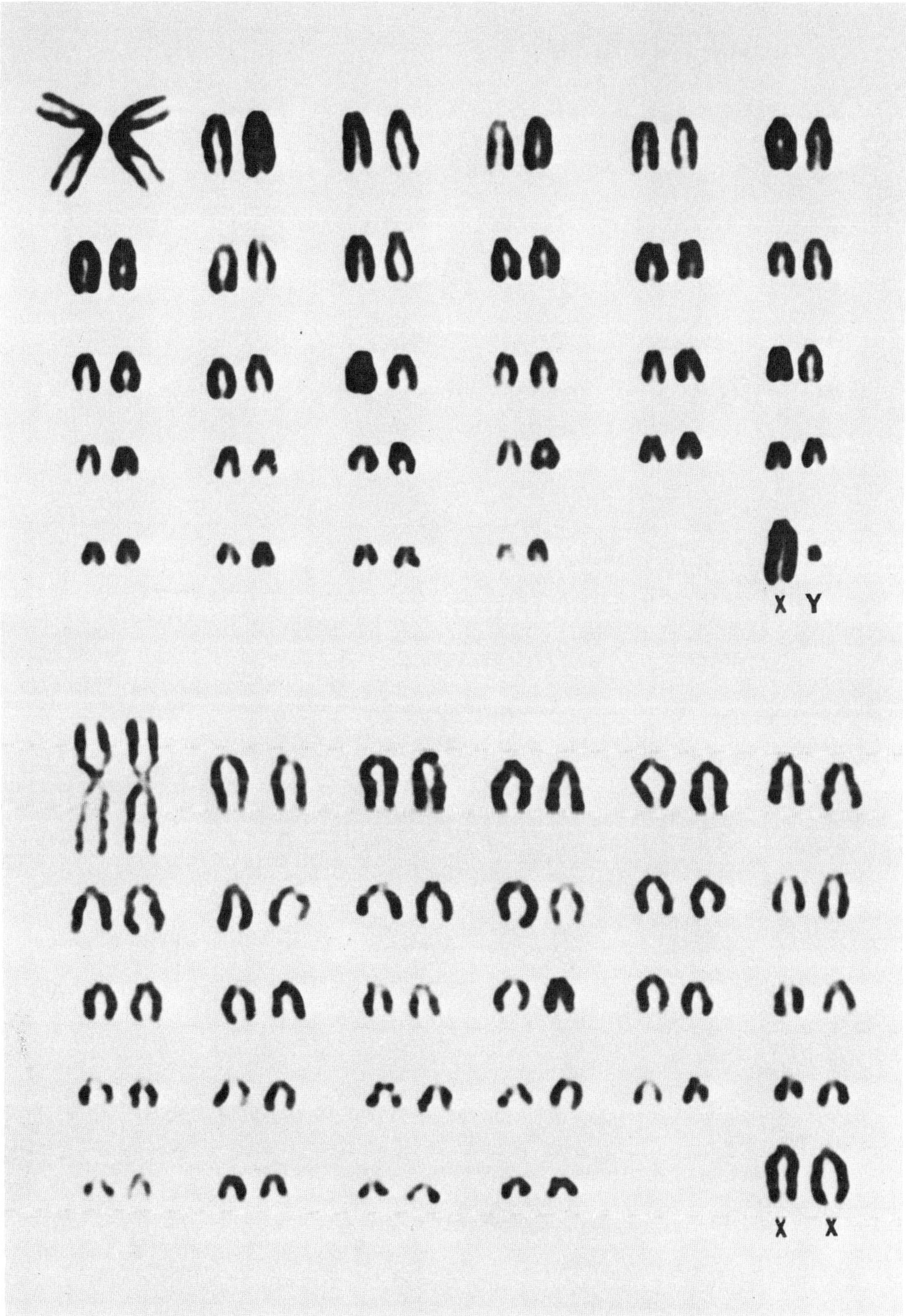

# Order: PRIMATES

***Lemur fulvus fulvus*** (Red-fronted lemur)

**Family: LEMURIDAE**

$2n = 48$

Order: PRIMATES Family: LEMURIDAE

<u>Lemur fulvus fulvus</u> (Red-fronted lemur)

2n=48

AUTOSOMES: 16 Metacentrics and submetacentrics
30 Acrocentrics

SEX CHROMOSOMES: X Acrocentric
Y Minute

Among the large metacentrics and submetacentrics roughly two groups of 4 pairs each can be recognized, one with the two arms more or less equal and the other distinctly unequal. One of the submetacentric pairs is unusually long. All the acrocentrics are relatively small, three pairs of which appear to bear secondary constriction near the centromere.

The sex chromosome can be unequivocally identified, the X being the longest acrocentric of the complement, and the Y, the smallest.

The karyotypes are gifts of Dr. Ernest H.Y. Chu, Biology Division, Oak Ridge National Laboratory, Oak Ridge, Tennessee, USA. The specimens were collected by Dr. John Buettner-Janusch in Madagascar.

REFERENCES:


1) Chu, E.H.Y. and Swomley, B.A.: Chromosomes of lemurine lemurs. Science 133:1925, 1961.

2) Bender, M.A. and Chu, E.H.Y.: The chromosomes of Primates. In "Evolutionary and Genetic Biology of Primates", Vol. 1 (J. Buettner-Janusch, ed.), Academic Press, N.Y., 1963.

3) Egozcue, J.: Chromosome variability in the Lemuridae. Amer. J. Phys. Anthrop. 26:341, 1967.

4) Egozcue, J.: Primates. In "Comparative Mammalian Cytogenetics" (Benirschke, K., ed.), Springer-Verlag, N.Y., 1969.

Order: PRIMATES

Family: LEMURIDAE

## *Lemur fulvus fulvus* (Red-fronted lemur)

2n=48

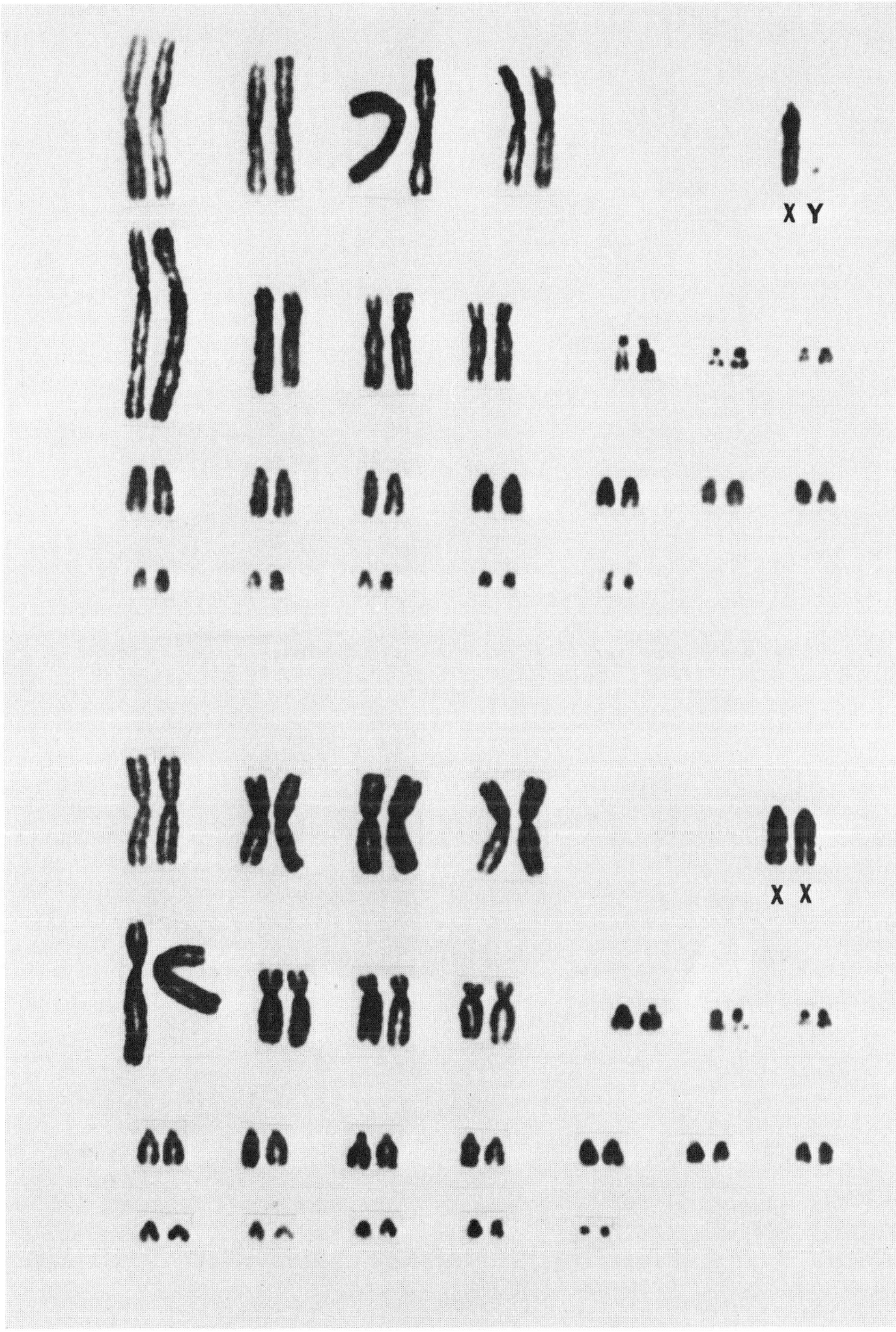

# Order: PRIMATES

*Leontocebus rosalia* (Golden marmoset)

## Family: CALLITHRICIDAE
## 2n = 46

*Leontocebus rosalia* (Golden marmoset)

2n=46

AUTOSOMES: 42 Metacentrics, submetacentrics and subtelocentrics
2 Acrocentrics

SEX CHROMOSOMES: X Submetacentric
Y Metacentric

The short arms of several autosomes are of sufficient length to place the elements into the group of subtelocentrics or submetacentrics; division is arbitrary and depends on the excellence of preparation. Bone marrows usually yield less favorable results. Male/female blood chimerism is found in many animals. The male karyotype came from a skin culture, the female karyotype from a bone marrow preparation. Only one pair is considered to be truly acrocentric and is placed here before the sex chromosomes.

REFERENCES:

1) Benirschke, K., Anderson, J.M. and Brownhill, L.E.: Marrow chimerism in marmosets. Science 138:513, 1962.

2) Benirschke, K. and Brownhill, L.E.: Further observations on marrow chimerism in marmosets. Cytogenetics 1:245, 1962.

3) Egozcue, J.: Primates. In *Comparative Mammalian Cytogenetics* (Benirschke, K., ed.), Springer-Verlag, New York, 1969.

4) Hsu, T.C. and Hampton, S.H.: Chromosomes of Callithricidae with special reference to an XX/'XO' sex chromosome system in *Callimico goeldii* (Thomas, 1904). Folia Primatologica (in press).

<u>Leontocebus rosalia</u> (Golden marmoset)

2n=46

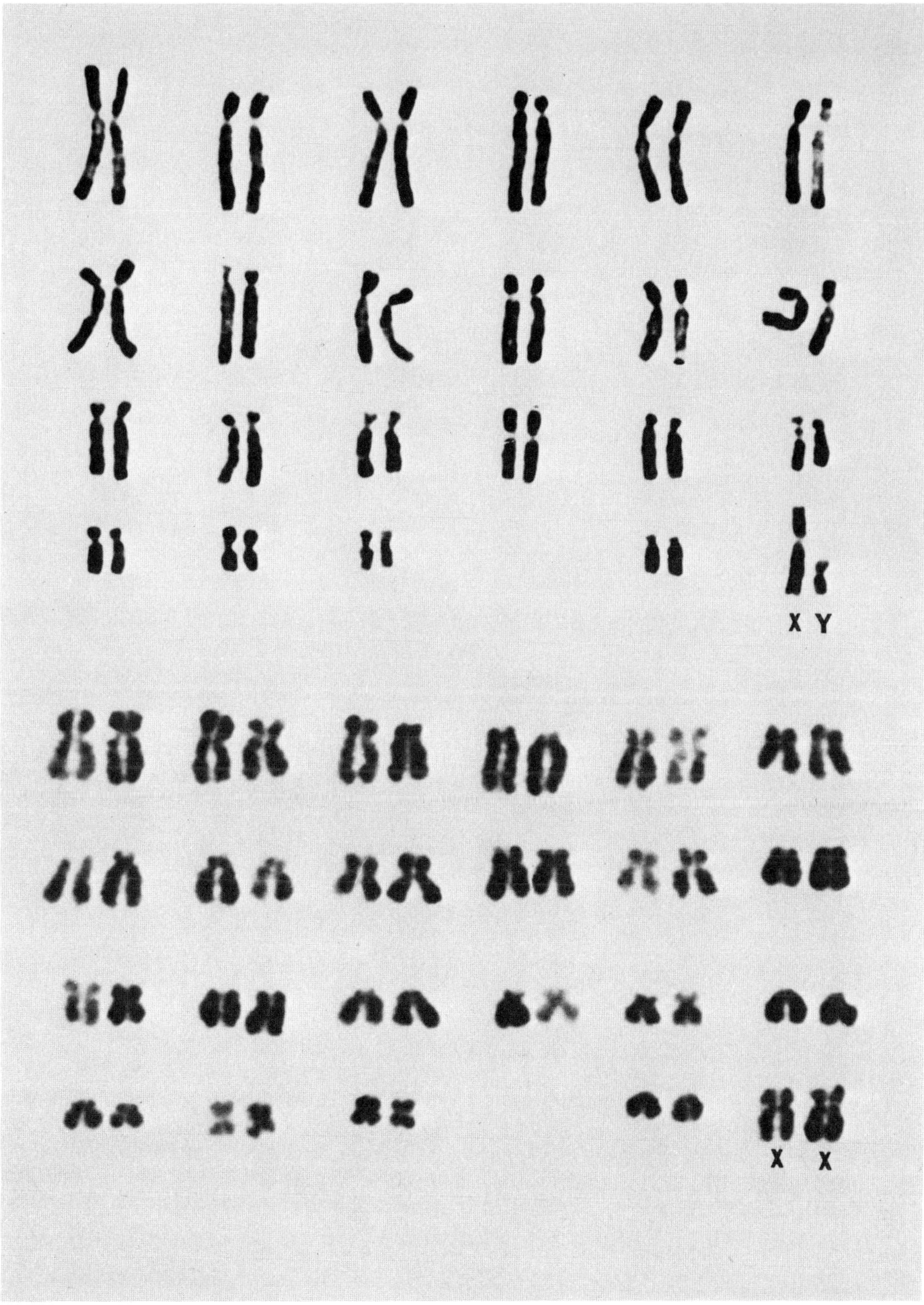

# Order: PRIMATES

## Family: CERCOPITHECIDAE

*Cercocebus torquatus* (Red crowned mangabey)
$2n = 42$

Order: PRIMATES Family: CERCOPITHECIDAE

*Cercocebus torquatus* (Red crowned mangabey)

2n=42

AUTOSOMES: 40 Metacentrics and submetacentrics

SEX CHROMOSOMES: X Submetacentric
Y Metacentric

Skin biopsies of a male and a female animal were kindly made available by Dr. C. Gray, National Zoological Park, Washington, D.C., USA. The Y chromosome is easily identified as the smallest element. The pair of marker chromosomes with secondary paracentromeric constrictions is placed as the last autosomes.

REFERENCES:


1) Chu, E.H.Y. and Giles, N.M.: A study of primate chromosome complements. Amer. Natur. 41:273, 1957.

2) Bender, M.A. and Mettler, L.E.: Chromosome studies of primates. Science 128:186, 1958.

3) Chiarelli, B.: Comparative morphometric analysis of the primate chromosomes. II. The chromosomes of the genera *Macaca*, *Papio*, *Theropithecus*, and *Cercocebus*. Caryologia 15:401, 1962.

4) Klinger, H.P.: The somatic chromosomes of some primates (*Tupaia glis*, *Nycticebus coucang*, *Tarsius bancanus*, *Cercocebus aterrimus*, *Symphalangus syndactylus*). Cytogenetics 2:140, 1963.

5) Chiarelli, B.: Marked chromosome in catarrhine monkeys. Folia primat. 4:74, 1966.

6) Chiarelli, B.: La morfologia del cromosoma "Y" delle differenti specie di primati. Rivista Antropol. 54:3, 1967.

7) Chiarelli, B. and Vaccarino, C.: Cariologia ed evoluzione nel genere *Cercopithecus*. Atti Ass. Genet. It., Pavia 9:328, 1964.

8) Chiarelli, B.: Caryology and taxonomy of the catarrhine monkeys. Amer. J. Phys. Anthropol. 24:155, 1966.

9) Wurster, D. and Benirschke, K.: Chromosomes of some primates. Mammalian Chromosomes Newsletter 10:3, 1969.

Cercocebus torquatus (Red crowned mangabey)

2n=42

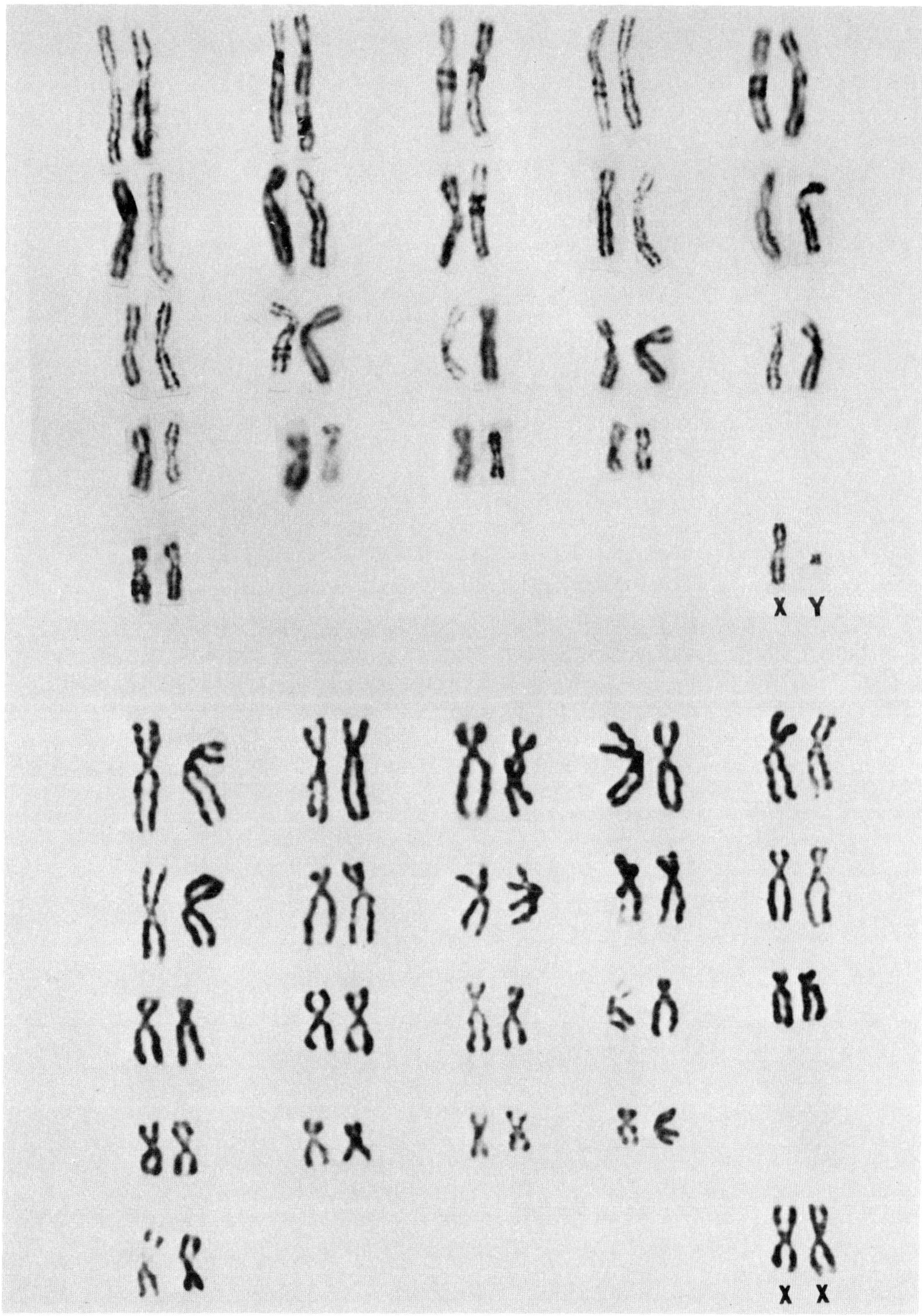

# Order: PRIMATES

## Family: CERCOPITHECIDAE

*Cercopithecus cephus* (Moustached guenon)
2n = 66

Order: PRIMATES Family: CERCOPITHECIDAE

*Cercopithecus cephus* (Moustached guenon)

2n=66

AUTOSOMES: 44 Metacentrics, submetacentrics and subtelocentrics
20 Acrocentrics

SEX CHROMOSOMES: X Submetacentric
Y Submetacentric

Some subtelocentric autosomes could easily be regarded as acrocentrics. The marker autosome is placed before the sex chromosomes. It is acrocentric and has secondary constrictions near the centromere. Skin biopsies of one male and three female animals yielded similar results. They were kindly provided by Dr. C. Gray, National Zoological Park, Washington, D.C., USA.

REFERENCES:


1) Wurster, D. and Benirschke, K.: Chromosomes of some primates. Mammalian Chromosomes Newsletter 10:3, 1969.

2) Chiarelli, B.: Primi resultati di richerche di genetica e cariologia comparata in primati e loro interese evolutivo. Rivista di Antropol. 50: 87, 1963.

3) Egozcue, J.: Primates. In *Comparative Mammalian Cytogenetics* (Benirschke, K., ed.), Springer-Verlag, New York, 1969.

4) Chiarelli, B.: Survey of the caryology of the genus *Cercopithecus*. Mammalian Chromosomes Newsletter No. 22, 209, 1966.

5) Chiarelli, B. and Vaccarino, C.: Cariologia ed evoluzione nel genere *Cercopithecus*. Atti Ass. Genet. It., Pavia 9:328, 1964.

6) Chiarelli, B.: Chromosome polymorphism in the species of the genus *Cercopithecus*. Cytologia 33:1, 1968.

<u>Cercopithecus</u> <u>cephus</u> (Moustached guenon)

2n=66

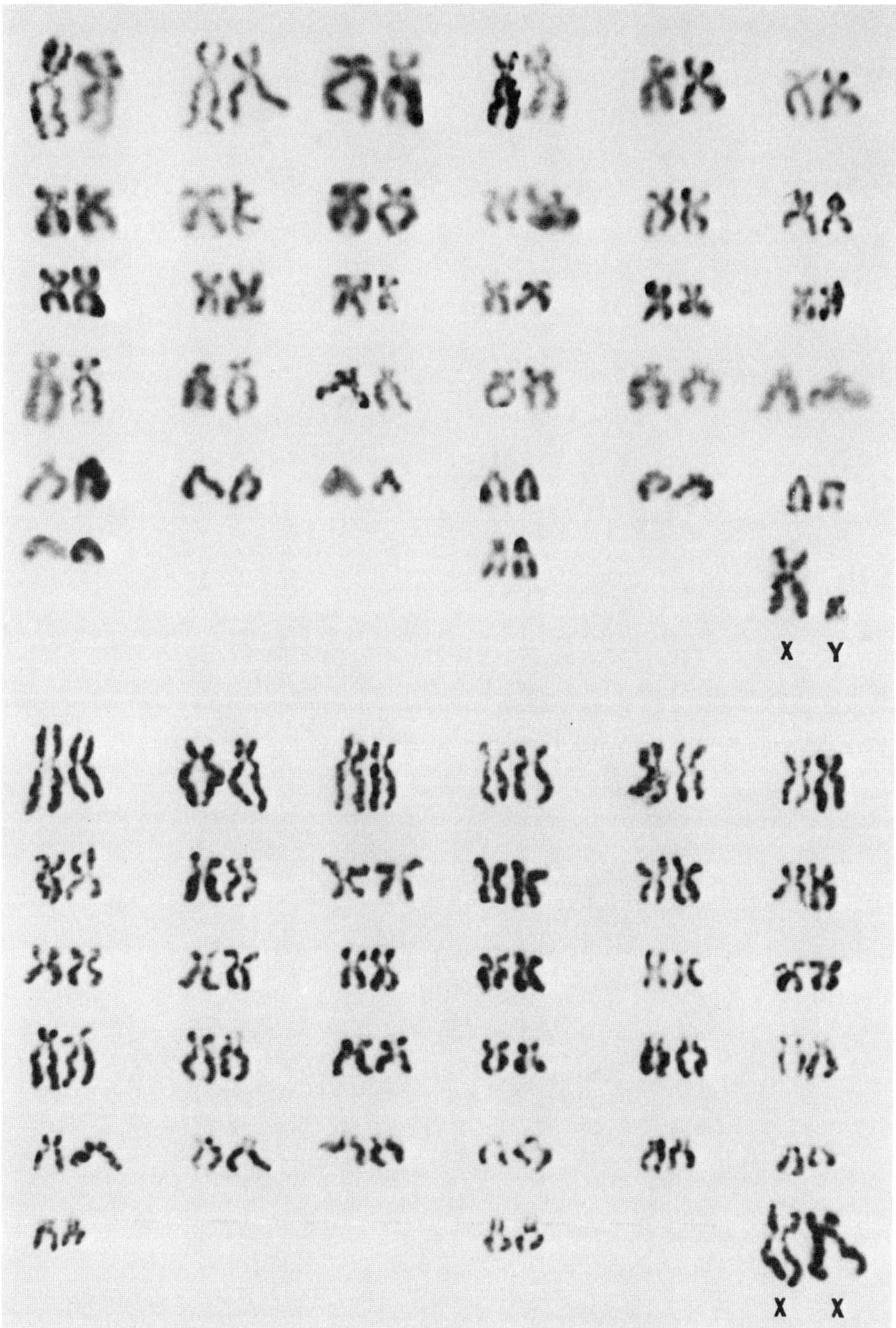

# Order: PRIMATES

## Family: CERCOPITHECIDAE

***Miopithecus (Cercopithecus) talapoin*** (Talapoin monkey; ozem)
$2n = 54$

Order: PRIMATES Family: CERCOPITHECIDAE

Miopithecus (Cercopithecus) talapoin (Talapoin monkey; ozem)

2n=54

AUTOSOMES: 38 Metacentrics and submetacentrics
14 Acrocentrics

SEX CHROMOSOMES: X Submetacentric
Y Submetacentric, minute

These karyotypes came from tissue cultures of skin biopsies and were donated by Dr. T. C. Jones, New England Regional Primate Research Center, Southborough, Massachusetts, USA. They were prepared in collaboration with Dr. R. Cooper, San Diego Zoo, California, USA. The arrangement into four groups of autosomes is that suggested by Dr. Jones. The acrocentrics shown above the sex chromosomes are the marker elements with a secondary constriction near the centromere. The Y has also been described as being acrocentric.

REFERENCES:

1) Chiarelli, B. and Vaccarino, C.: Cariologia ed evoluzione nel genere Cercopithecus. Atti Ass. Genet. It., Pavia 9:328, 1964.

2) Chiarelli, B.: Caryology and taxonomy of the catarrhine monkeys. Amer. J. Phys. Anthrop. 24:155, 1966.

3) Barberis, L.: The idiograms of three species of the genus Cercopithecus, C. talapoin, C. aethiops and C. mona. In Progress in Primatology (Starck, D., Schneider, R. and Kuhn, H.J., eds.), Gustav Fischer, Stuttgart, 1967.

4) Chiarelli, B.: Chromosome polymorphism in the species of the genus Cercopithecus. Cytologia 33:1, 1968.

5) Chiarelli, B.: Some new data on the chromosomes of Catarrhina. Experientia 18:405, 1962.

6) Egozcue, J.: Primates. In Comparative Mammalian Cytogenetics (Benirschke, K., ed.), Springer-Verlag, New York, 1969.

Miopithecus (Cercopithecus) talapoin (Talapoin monkey; ozem)

2n=54

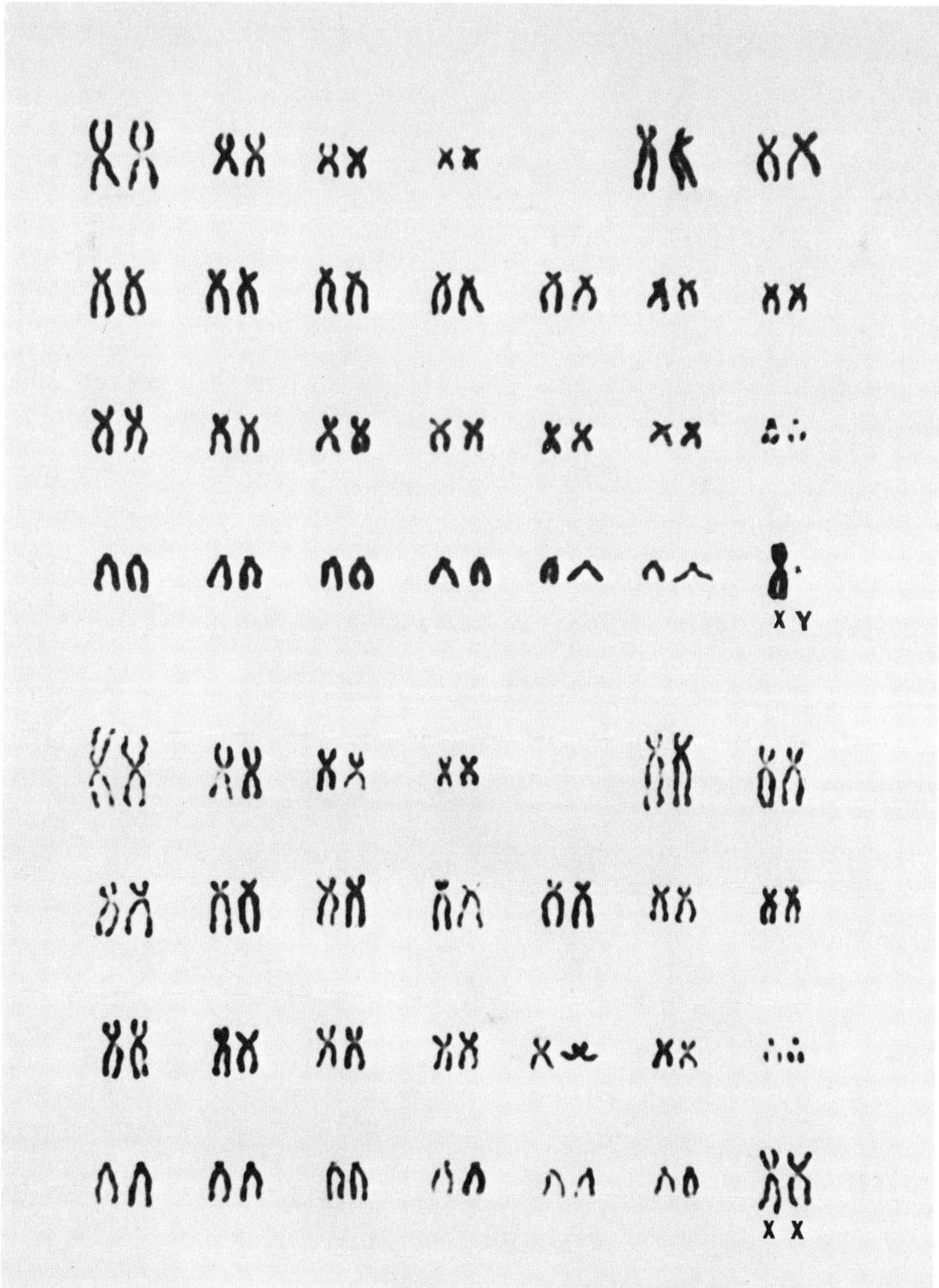

# Order: PRIMATES

## Family: CERCOPITHECIDAE

***Presbytis obscurus*** (Dusky langur)

2n = 48

Order: PRIMATES Family: CERCOPITHECIDAE

*Presbytis obscurus* (Dusky langur)

2n=44

AUTOSOMES: 40 Metacentrics and submetacentrics
2 Acrocentrics

SEX CHROMOSOMES: X Submetacentric
Y Submetacentric

Skin biopsies of a male and female animal were kindly made available by Dr. C. Gray, National Zoological Park, Washington, D.C., USA. The karyotype has previously (1) been described erroneously as belonging to *P. phayrei* and we are grateful to Dr. C. A. Hill, San Diego Zoo, California, USA., for clarifying the error.

Pairing of autosomes is relatively arbitrary except for the smallest metacentrics, the acrocentric pair, and the marker pair which is placed as the last autosomal pair. At least one metacentric has very small satellites.

REFERENCES:

1) Wurster, D. and Benirschke, K.: Chromosomes of some primates. Mammalian Chromosomes Newsletter 10:3, 1969.

2) Chiarelli, B.: Cariologia ed evoluzione nel genere *Cercopithecus*. Atti Ass. Genet. It., Pavia 9:328, 1964.

3) Chiarelli, B.: Caryology and taxonomy of the catarrhine monkeys. Amer. J. Phys. Anthropol. 24:155, 1966.

4) Chiarelli, B.: La morfologia del cromosome "Y" delle differenti specie di primati. Rivista Antropol. 54:3, 1967.

5) Chiarelli, B.: Comparative morphometric analysis of primate chromosomes. III. The chromosomes of the genera *Hylobates*, *Colobus* and *Presbytis*. Caryologia 16:637, 1963.

*Presbytis obscurus* (Dusky langur)

2n=44

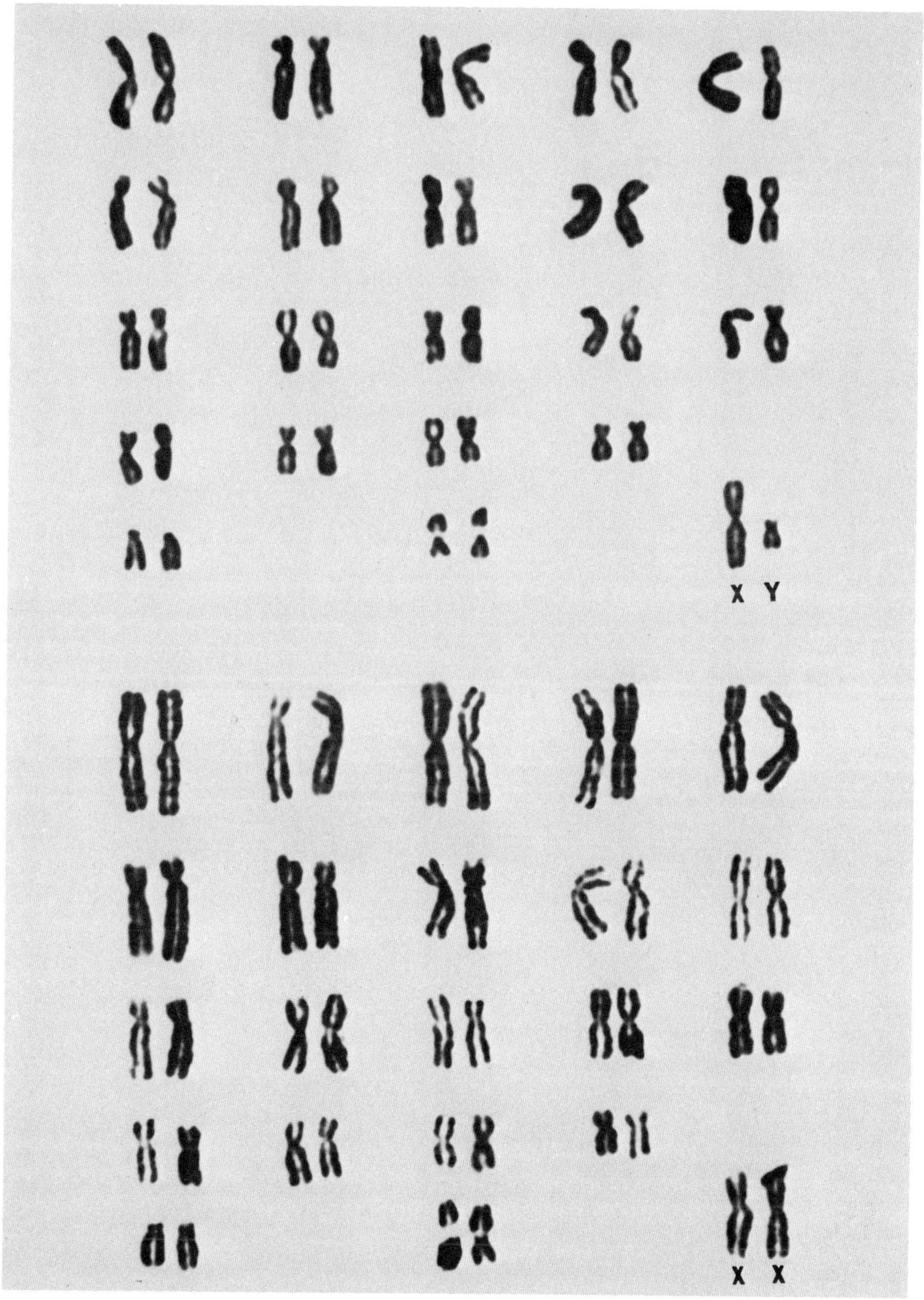

# Order: PRIMATES

## Family: PONGIDAE

***Pongo pygmaeus*** (Orangutan)
$2n = 44$

Order: PRIMATES Family: PONGIDAE

*Pongo pygmaeus* (Orangutan)

2n=48

AUTOSOMES: 28 Metacentrics or submetacentrics
18 Acrocentrics

SEX CHROMOSOMES: X Submetacentric
Y Acrocentric or submetacentric

Skin biopsies of one male and one female were kindly made available by Dr. C. Gray, National Zoological Park, Washington, D.C., USA. Lymphocyte cultures were obtained of the pair of fraternal twins at the Zoo in Seattle, Washington, USA, through the courtesy of Dr. D. Smith. Skin biopsy of a male newborn came from the San Diego Zoo. All gave similar results; no chimerism was found in the twins. The karyotypes displayed here came from skin cultures, the male from "Archie", the female from "Susie" at Washington, D.C. Faint satellites can be seen on some acrocentrics.

REFERENCES:

1) Chiarelli, B.: Chromosomes of the Orang-Utan (*Pongo pygmaeus*). Nature 192:825, 1961.

2) Hamerton, J.L., Klinger, H.P., Mutton, D.E. and Lang, E.M.: The somatic chromosomes of the Hominoidea. Cytogenetics 2:240, 1963.

3) Klinger, H.P., Hamerton, J.L., Mutton, D. and Lang, E.M.: The chromosomes of the Hominoidea. In *Classification and Human Evolution* (Washburn, S.L., ed.), Aldine Publishing Co., Chicago, 1963.

4) Egozcue, J.: Primates. In *Comparative Mammalian Cytogenetics* (Benirschke, K., ed.), Springer-Verlag, New York, 1969.

5) Chiarelli, B.: Comparative morphometric analysis of the primate chromosomes. I. The chromosomes of the anthropoid apes and of man. Caryologia 15:99, 1962.

6) Chiarelli, B.: La morfologia del cromosoma "Y" delle differenti specie di primati. Rivista Antropol. 54:3, 1967.

Order: PRIMATES Family: PONGIDAE

*Pongo pygmaeus* (Orangutan)

2n=48

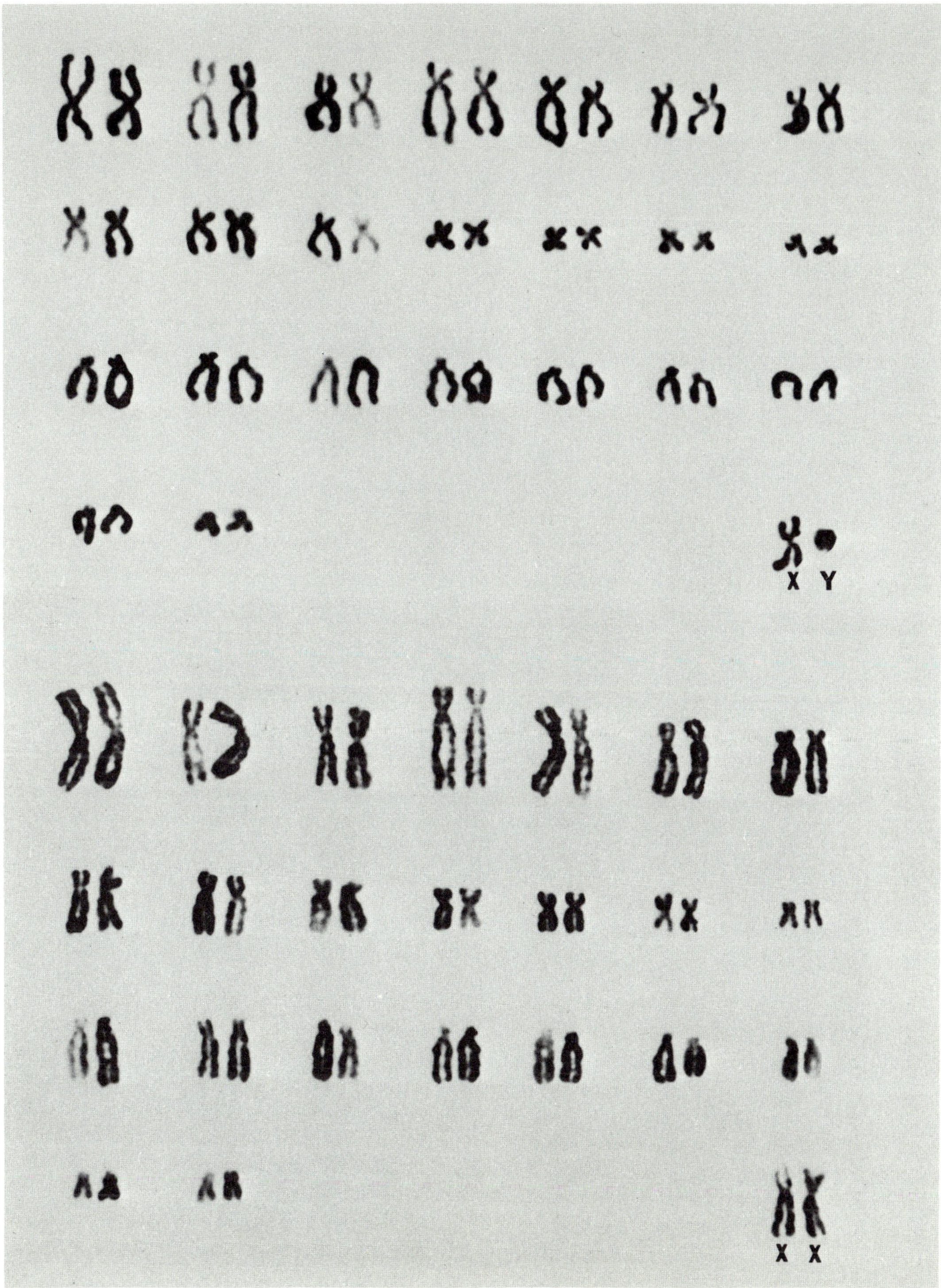

# Cumulative Index (Volumes 1 to 3)

## Vernacular Names

# Cumulative Index (Volumes 1 to 3)

## Scientific Names